Addressing Concerns About Tastes and Odors and Cyanotoxins in Tap Water

AF849309

Awwa Research Foundation

About the Awwa Research Foundation

The Awwa Research Foundation (AwwaRF) is a member-supported, international, nonprofit organization that sponsors research to enable water utilities, public health agencies, and other professionals to provide safe and affordable drinking water to consumers.

The Foundation's mission is to advance the science of water to improve the quality of life. To achieve this mission, the Foundation sponsors studies on all aspects of drinking water, including supply and resources, treatment, monitoring and analysis, distribution, management, and health effects. Funding for research is provided primarily by subscription payments from approximately 1,000 utilities, consulting firms, and manufacturers in North America and abroad. Additional funding comes from collaborative partnerships with other national and international organizations, allowing for resources to be leveraged, expertise to be shared, and broad-based knowledge to be developed and disseminated. Government funding serves as a third source of research dollars.

From its headquarters in Denver, Colorado, the Foundation's staff directs and supports the efforts of more than 800 volunteers who serve on the board of trustees and various committees. These volunteers represent many facets of the water industry, and contribute their expertise to select and monitor research studies that benefit the entire drinking water community.

The results of research are disseminated through a number of channels, including reports, the Web site, conferences, and periodicals.

For subscribers, the Foundation serves as a cooperative program in which water suppliers unite to pool their resources. By applying Foundation research findings, these water suppliers can save substantial costs and stay on the leading edge of drinking water science and technology. Since its inception, AwwaRF has supplied the water community with more than $300 million in applied research.

More information about the Foundation and how to become a subscriber is available on the Web at **www.awwarf.org.**

Addressing Concerns About Tastes and Odors and Cyanotoxins in Tap Water

Prepared by:
Randy Gottler and **Jennifer Calles**
City of Phoenix Water Services Department
Laboratory Services
and
Paul Westerhoff and **Morteza Abbaszadegan**
Department of Civil and Environmental Engineering, Arizona State University
Tempe, AZ 85287-5306

Jointly sponsored by:
Awwa Research Foundation
6666 West Quincy Avenue, Denver, CO 80235-3098
and
City of Phoenix Water Services Department

Published by:

Awwa Research Foundation

American Water Works Association

IWA Publishing

DISCLAIMER

This study was jointly funded by the Awwa Research Foundation (AwwaRF) and the City of Phoenix Water Services Department (Phoenix). AwwaRF and Phoenix assumes no responsibility for the content of the research study reported in this publication or for the opinions or statements of fact expressed in the report. The mention of trade names for commercial products does not represent or imply the approval or endorsement of AwwaRF or Phoenix. This report is presented solely for informational purposes.

Copyright © 2007
by Awwa Research Foundation

All Rights Reserved

CONTENTS

TABLES

FIGURES

FOREWORD

The Awwa Research Foundation is a nonprofit corporation that is dedicated to the implementation of a research effort to help utilities respond to regulatory requirements and traditional high-priority concerns of the industry. The research agenda is developed through a process of consultation with subscribers and drinking water professionals. Under the umbrella of a Strategic Research Plan, the Research Advisory Council prioritizes the suggested projects based upon current and future needs, applicability, and past work; the recommendations are forwarded to the Board of Trustees for final selection. The foundation also sponsors research projects through the unsolicited proposals process; the Collaborative Research, Research Applications, and Tailored Collaboration programs; and various joint research efforts with organizations such as the U.S. Environmental Protection Agency, the U.S. Bureau of Reclamation, and the Association of California Water Agencies.

This publication is a result of one of these studies, and it is hoped that its findings will be applied in communities throughout the world. The following report serves not only as a means of communicating the results of the water industry's centralized research program but also as a tool to enlist the further support of the nonmember utilities and individuals.

Projects are managed closely from their inception to the final report by the foundation's staff and large cadre of volunteers who willingly contribute their time and expertise. The foundation serves a planning and management function and awards contracts to other institutions such as water utilities, universities, and engineering firms. The funding for this research effort comes primarily from the Subscription Program, through which water utilities subscribe to the research program and make an annual payment proportionate to the volume of water they deliver and consultants and manufacturers subscribe based on their annual billings. The program offers a cost-effective and fair method for funding research in the public interest.

A broad spectrum of water supply issues is addressed by the foundation's research agenda: resources, treatment and operations, distribution and storage, water quality and analysis, toxicology economics, and management. The ultimate purpose of the coordinated effort is to assist water suppliers to provide the highest possible quality of water economically and reliably. The true benefits are realized when the results are implemented at the utility level. The foundation's trustees are pleased to offer this publication as a contribution toward that end:

Walter J. Bishop
Chair, Board of Trustees
Awwa Research Foundation

Robert C. Renner, P.E.
Executive Director
Awwa Research Foundation

ACKNOWLEDGMENTS

The authors of this report are indebted to the following water utilities and other individuals for their support, cooperation and participation in this project:

City of Tempe, Arizona
City of Scottsdale, Arizona
City of Phoenix Water Services Department, Pollution Control Division, Phoenix, AZ – Steve Schoen, Debbie Sue Estrada, and Michelle Heilman. Laboratory Services Division – Terry Kitchen, Philip Johnson, Dorothy Robinson and Joannie Knigge.
Arizona State University Grad students – Michelle Cummings, Min Yi, Anthony Stevens, Alby Aguilar, and Mario Esparza-Soto.

The authors also wish to acknowledge the input and guidance provided by AwwaRF Project Manager Djanette Khiari, and the Project Advisory Committee (PAC) – including Andrea Dietrich, Virginia Tech, Mel Suffet, UCLA, and Marcia Torobin, Metropolitan Water District of Southern California.

EXECUTIVE SUMMARY

RESEARCH OBJECTIVES

The goal of this project is to develop a monitoring and response program to rapidly assess emerging problems in the distribution system, and provide security to customers that unpleasant tastes and odors are not indicators of unsafe water. The specific objectives of this project are as follows:

1. Quantify non-methyl isoborneol (MIB)/Geosmin Taste and Odor (T&O) compounds from waters and cultures.
2. Streamline FPA and cyanotoxin methods for routine laboratory monitoring.
3. Investigate biological production of T&O in the distribution systems.
4. Develop and demonstrate a responsive T&O distribution system monitoring program.

APPROACH

The project was divided into the following tasks, which are detailed in subsequent sections:

Task 1 – Develop a Flavor Profile Analysis (FPA) Group
The City of Phoenix established two FPA panels trained in identifying and characterizing the T&O of the water.

Task 2 – Quantify non-MIB/Geosmin Tastes and Odors
Identification of specific compounds more accurately identified the cause of T&O episodes and facilitated appropriate treatment.

Task 3 – Develop a Capability for Analyzing Algal Biotoxins
The presence of cyanobacteria (blue-green) algae in the water supply reservoir has contributed to significant T&O problems.

Task 4 – Conduct Baseline Water Supply Monitoring for Algal Biotoxins
Reservoir samples were analyzed monthly for microcystin.

Task 5 – Conduct Baseline Water Distribution System Monitoring
Twenty sampling points per month were monitored in the distribution system for pH, temperature, total dissolved solids (TDS), nitrate, chlorine residual, bacterial numeration, FPA and T&O. Correlations were determined between FPA and T&O data.

Task 6 – Conduct Responsive Water Distribution Systems Monitoring
Distribution system samples were monitored based on customer complaints. Site visit procedures and complaint inquiry forms were developed and implemented. An intensive sampling effort in a historically problematic area was conducted to determine if distribution system formation of T&O occurred.

Task 7 – Culture of Water Distribution System Organisms
Odor-producing organisms were isolated from distribution system samples collected in the baseline monitoring effort.

Task 8 – Conduct Laboratory Experiments with Culprit Water Distribution System

Organisms – Stimulation and Control
Pipeloop experiments were used to determine depletion rates and concentrations of MIB, Geosmin and actinomycetes.

Task 9 – Conduct Field Control Culprit Organisms
T&O production and colonization by actinomycetes were assessed in a simulated distribution system. The effect of stagnation versus flowing water; chlorination; and surface versus ground water were compared.

CONCLUSIONS

1. Microcystin concentration in the Phoenix water supply varies with time.
2. Microcystin peak production precedes MIB peak concentrations by one month.
3. β-cyclocitral can be produced by algae and may play significant role in T&O occurrence.
4. Powder activated carbon (PAC) used for T&O removal absorbs Microcystin.
5. Drinking water distribution systems can contribute to T&O problems of tap water.
6. Earthy/Musty descriptors showed a direct correlation with MIB and geosmin but a weak correlation with β-cyclocitral.
7. Earthy/Musty FPA intensity could be predicted with the MIB and geosmin plots and with a three component (MIB, Geosmin, β-cyclocitral) plot.
8. FPA panels were difficult to staff and results were variable.
9. Distribution system production of tastes and odors did not occur or could not be detected.
10. Customer complaints can increase when chlorine residuals are low. High chlorine concentration can mask tastes and odors.
11. Changes in source water can provide favorable conditions for MIB production.

RECOMMENDATIONS FOR UTILITIES

1. Implement use of powdered activated carbon (PAC) addition.
2. Address seasonal T&O events before it becomes a significant problem, by analyzing for tastes and odors and utilize results to predict customer complaints.
3. Coordinate response to customer complaints by obtaining more specific information.
4. Monitor production trends to determine critical production periods throughout the year.
5. Maintain chlorine residuals and minimize stagnant water in the distribution system.
6. Monitor Microcystin based on dynamic nature of cyanobacteria blooms.
7. Gas Chromatograph/Mass Spectrometry Solid Phase Micro Extraction (SPME) is recommended for accurate and rapid T&O analyses.
8. Enzyme-linked ImmunoSorbent assays (ELISA) are recommended for microcystin analyses.
9. Development of better biofilm sampling protocols.

CHAPTER 1
INTRODUCTION AND OBJECTIVES

STATEMENT OF PROBLEM

Consumers today demand drinking water that looks, smells, and tastes good, in addition to meeting guidelines for regulated constituents. The occurrence of aesthetically unpleasant T&O compounds in water creates a perception of unsafe water and frequently results in consumer complaints or leads consumers to alternate drinking water sources (bottled water, point of use devices). The main sources of earthy/musty T&O problems in municipal water supplies are certain types of algae, mainly blue-green algae (cyanobacteria) and actinomycetes. Although many organic compounds may impart tastes and odors to water, by far the most commonly reported are two chemicals: geosmin and 2-methylisoborneol (MIB). However, focusing on just these two compounds, which are analyzed by Gas Chromatograph/Mass Spectrometry (GC/MS), may preclude understanding the presence and treatment needs of other compounds noticeable to consumers.

Algae related T&O problems are common among municipalities that use surface water. The problem is most common during the summer and fall. Although in-plant treatments (e.g., ozone, activated carbon) are effective at reducing T&O levels, it is generally more cost effective to prevent T&O problems at the source (Suffet et al. 1993; Suffet et al. 1995). However, in-plant treatments are ineffective if taste or odors are produced by culprit organisms in the finished-water distribution system (Khiari et al. 2000).

At least 25 types of algae are confirmed to produce geosmin and about a dozen produce MIB. Both compounds create detectable odor at extremely low concentrations (5 or 10 ng/L for MIB and 10 ng/L for geosmin; Wnorowski, 1992). Other compounds can produce tastes and odors, but these have been less well studied and are apparently less common. T&O problems have typically been associated with phytoplankton (free-floating algae; Izaguirre et al. 1982; Hayes and Burch, 1989). However, research by the Metropolitan Water District of Southern California (MWDSC) has shown that T&O problems can also be caused by periphyton (attached algae). Periphyton are the main source of T&O problems in Lake Matthews, California (Means and McGuire, 1986) and in the California Aqueduct (Izaguirre and Taylor, 1995) systems. T&O producing periphyton in the California Aqueduct were found in small patches, generally during reduced flow.

Municipalities in the Phoenix metropolitan area have long experienced T&O problems. Most of the water used by municipalities in the Phoenix metropolitan area is supplied by the Salt and Verde Rivers, Salt River Project (SRP) and ground water and/or Central Arizona Project (CAP) systems. Colorado River water is also conveyed to the Tucson area. Both the CAP and SRP systems utilize storage reservoirs and both have extensive canal systems that convey water to water treatment plants. Both the reservoirs and the canal systems are potential sources of T&O compounds. All municipalities in the SRP service area also have T&O problems. One city, for example, now spends over $1 million per year to add powdered activated carbon (PAC) to their water treatment plants in an attempt to reduce T&O problems, with only limited success.

The Cities of Phoenix (COP), Tempe, and Scottsdale, Arizona have recognized the need to develop a long-term implementation strategy to respond to customer complaints regarding these nuisance tastes and odors.

PROJECT OBJECTIVES

The goal of this project is to develop a monitoring and response program to rapidly assess emerging problems in the distribution system, and provide security to customers that nuisance tastes and odors are not indicators of unsafe water.
The specific objectives of this project are as follows:

- Quantify non-MIB/Geosmin T&O compounds from waters and cultures
- Streamline cyanotoxin methods for routine laboratory monitoring
- Investigate biological production of T&O compounds in the distribution system.
- Develop and demonstrate a responsive T&O distribution system monitoring program.

The remainder of this report is divided into seven chapters. Chapters 2 and 3 describe the sampling sites, laboratory analytical methods and intracellular extraction. Chapter 4 focuses on the effect of seasonal changes, PAC, and chlorine dosage on microcystin concentrations. In Chapter 5, the sampling points and results for the study period are discussed. Various calculations are explored to correlate FPA with T&O concentrations. Chapter 6 provides detail of laboratory studies relating to inactivation of Actinomycetes. Chapter 7 examines the effect of biofilms within a small scale distribution system. Chapter 8 summarizes the key findings of the research and offers recommendations to utilities.

CHAPTER 2
MATERIALS AND METHODS

This chapter describes the locations from which samples were collected, sampling methods, experimental methods for laboratory tests, and analytical methods.

ANALYTICAL METHODS

Flavor Profile Analysis Group

The City of Phoenix established two Flavor Profile Analysis panels (FPA) trained in identifying and characterizing the T&O of the water. Twenty distribution sites were sampled each month for analysis by FPA. The sampling was conducted for 18 months. One-liter samples were collected in amber glass containers for odor. A 250 mL amber glass bottle was collected for taste. When sampling from a tap the system was allowed to flush for five minutes. The flow was adjusted to about 500 mL/min and the bottle was rinsed with the sample. The sample was collected with no headspace. All samples were iced at 4° C immediately after collection. The samples were analyzed within 48 hours of collection.

Taste standards were provided as reference points for both T&O intensities. Sweet standards were made at the following concentrations:

- 25 grams sugar to 500 mL water = intensity rating 4
- 50 grams sugar to 500 mL water = intensity rating 8
- 75 grams sugar to 500 mL water = intensity rating 12

A 30 ng/L standard was utilized as a reference for MIB and Geosmin. For both T&O analysis, each panelist determined intensity ratings for each sample. After each panel member individually rated the samples a panel discussion was held. All results for both T&O were recorded by sample, panelist intensity rating, and average for the panel for each T&O descriptor. The strength of a taste or odor is judged according to the following scale: - (Odor-Free); T (Threshold); 2 (Very Weak); 4 (Weak); 6; 8 (Moderate); 10; and, 12 (Strong). Numerical ratings from each panelist were used to calculate averages for each T&O descriptor.

The City of Phoenix increased residual chlorine levels in response to the events of September 11, 2001. Since that time the panels experienced difficulties positively identifying specific odors. This was most likely due to dominant chlorine odors which were masking other odors causing fatigue in the panel members. To investigate the odor fatigue the samples were dechlorinated prior to testing by the FPA. In July 2002 FPA leads implemented dechlorination of samples with Hydrogen Peroxide. To dechlorinate, FPA leads added 100ul of a 3% hydrogen peroxide solution to 200 ml of sample. The dechlorinated samples produced a better correlation between the FPA panel results and the GC/MS results.

The panels continued to evaluate water treatment plant samples in an effort to provide early warning and near real time information on T&O levels for water treatment plant operators. For T&O control the operators used the results to increase or decrease PAC addition.

The City of Phoenix FPA results were evaluated for effectiveness of the FPA versus direct MIB and Geosmin analysis by Gas Chromatography/Mass Spectroscopy (GC/MS). Additionally, details were distributed to FPA team leaders and another municipality.

GC/MS Analysis for Taste-and-Odor Compounds

MIB and geosmin stock solutions in methanol were purchased from Sigma Chemical Company (MO, US). MIB and geosmin neat chemicals were purchased from Wako Chemicals Inc. All other chemicals used in laboratory experiments were reagent grade or better.

MIB and geosmin were quantified by Solid-Phase Micro-extraction method using Gas Chromatography with Mass Spectroscopy (SPME-GC/MS) (Lloyd et al 1998, Watson et al 2000). Briefly, 12 mL of sample was added to a 20 mL septum capped vial that contains 4 gm desiccated sodium chloride (NaCl). An auto sampler was used instead of manually pre-treating and injecting the samples. The prepared 20 mL capped sample vial was put on the auto sampler tray, then brought automatically to a magnetic heating and shaking cell for 6 minutes at 40° C. A SPME fiber (Supelco # 57348 U) was automatically introduced into the headspace through the septum of the sample vial in the heating cell, and shaken for 30 minutes. Then the fiber was removed from the vial and inserted into the gas chromatograph injector at 250° C for 6 minutes. The fiber was then retracted into the holder, removed from the GC inlet and back to the standby position waiting for reuse for the next sample. Compounds were eluted from the column gas chromatograph to a mass spectrometer set for selective ion storage (selective mass to charge ratio, m/z values: MIB = 95, geosmin = 112 and 2-Isopropyl-3-Methoxypyrazine (IPMP) = 124, 136). Calibration curves were generated using MIB and geosmin standards that were prepared in the lab. Standards ranged from 2 ppt to 100 ppt, which represent the detected typical MIB and geosmin concentration range. Figure 2.1 presents an example standard curve for MIB and geosmin.

The method detection limit (MDL) for MIB and geosmin was 2 ng/L. The detection sensitivity was ± 15%, which was decided by the standard deviation when running the same standard samples in replicates. The SPME method for MIB and geosmin analysis applied in this study used external standards for quality control (QC). A calibration curve was generated first when starting a new cycle of MIB and geosmin analysis. A QC standard (prepared in lab from the purchased standard solutions from Wako Inc.) was analyzed in between every 8~15 samples depending on the stability characteristics of the GC/MS instrument. All QC standards used were at a concentration of 20 ppt. The calibration curve was renewed whenever the QC standards were beyond 15% difference of the standard curve.

The method was expanded to include cis-3-hexanen-ol (MDL 32 ng/L), cis-3-hexenyl acetate (MDL 3.1 ng/L), 2-isobutyl-3-methoxypyrazine (IBMP) (MDL 1.4 ng/L), 2-isopropyl-3-methoxypyrazine (IPMP) (MDL 1.8 ng/L), and β-cyclocitral (MDL 1.9 ng/L). Calibration standards were prepared from Supelco neats at concentrations in water 2, 5, 20, 35, and 50 ng/L. Cis-3-hexen-1-ol was analyzed at a concentration 10 times than those listed.

Cyanotoxin Analysis

Microcystin Analysis was conducted using ELISA methods. One task of this project focused on adopting existing techniques developed for cyanotoxin analysis of algae blooms in lakes to rapid analysis of drinking water samples. Chapter 3 includes details of method development validation.

Actinomycete Analysis

Actinomycete counts were determined by the standard cultural method. Samples were assayed for actinomycetes using a double layer starch-casein agar (Standard Methods, 20th ed. # 9250). For the top agar, 4 mL of starch-casein media was added to a 16 x 150 mm glass tube and stored at 4° C. The bottom agar plates were prepared by adding 12 mL of starch-casein agar to a 15 x 100 mm plate (VWR). The agar was then allowed to solidify and stored inverted at 4° C.

Prior to sampling, the top agar was melted and then sustained at 48° C in a VWR heat block (West Chester, PA). In order to suppress fungal growth 0.5 mL of cycloheximide from a stock solution of 1 mg/mL was added to the top agar before adding the water sample (Sigma, St. Louis, MO). After the addition of cycloheximide, 1 mL of the water sample was added to the top agar tube, lightly vortexed, and then poured onto the bottom agar. Each plate was then allowed to set on the table until the media solidified. The plates were then parafilmed to reduce dehydration and stored in an inverted position for seven days at 28° C (Pechiney Plastic Packaging, Chicago, IL).

After seven days of incubation, the actinomycetes colonies were counted. Per Standard Methods protocol, actinomycetes colonies were multiplied by two, because the colonies often group together in small clusters and cannot be differentiated during counting, therefore the multiplication accounts for this phenomenon.

Biofilm Analysis

Despite attempts to obtain biofilms from the City of Phoenix water distribution system, no good cultures could be isolated. Additional activities were coordinated through the Water Distribution Division for the City of Phoenix. In May 2002 the laboratory received sections of pipe removed from the distribution system as part of a valve replacement job. The pipe was submerged in water and transported to the lab. When received by the lab, staff learned that the pipe section had been stored in this water for over two days before its delivery. Nonetheless, the pipe section was examined and photographed. Initial examination of the pipe showed no visible biofilm growth, pipe corrosion or pipe tuberculation. The interior pipe surface was scraped and the material was examined under a microscope. No biofilm organisms were observed.

Because of the length of time between removal from the distribution system and delivery to the laboratory, staff sought alternate approaches to obtaining samples. Discussions with Water Distribution personnel led to a recommendation that samples be obtained from valve tapping crews. The crews remove four to six inch pipe sections from the distribution system whenever new valves are placed into service. Using this process, disturbances to the inside surface of the pipe were minimal. Coupons were placed into sterile bags and delivered to the laboratory on the day they were removed from the distribution system.

Two coupons removed from concrete pipe at different locations were collected on June 26, 2002. The coupons were transported directly to the lab, examined and photographed. The examination again revealed no indication of biofilms. At that time the City of Phoenix maintained a 2.5 mgCl_2/L residual leaving the WTPs beginning in the fall of 2001 which is believed to have decreased biofilm occurrence.

EXPERIMENTAL METHODS

Pilot Distribution System Pipeloop Experimental Setup

Two independent laboratory scale pipeloops were constructed using cast-iron and PVC pipe (Figure 2.2). Each pipeloop system had a two-inch diameter distribution main of eighteen feet in length connected to a reservoir. The systems consisted of six removable pipe sections and a four-foot long dead end line with an additional two removable sections. Seven different sampling ports (including reservoir) were built into each pipeloop to assess the impacts of various tap configurations. Three representative sampling ports were selected for this study: Port#7 (Reservoir/ Influent), Port#1 (Gate valve, unthreaded type, soldered), and Port#3 (Dedicated sampling tap at the terminus of the dead end line).

The total volume contained in each pipeloop system was thirty liters of tap water. The water was continuously recirculated in the main distribution line and reservoir using a self-priming, thermally protected magnetic-drive electric pump (Little Giant Pump Company, Oklahoma City, OK). A rotameter and inline ball valve was installed for controlling the flow at 1 foot per second throughout the project. The temperature in each pipeloop was maintained at 20° C – 26° C.

This research focused on the effect of operational and non-operational changes on the survivability and metabolite (MIB and Geosmin) production of actinomycetes. The specific actinomycetes strain used in these experiments was *Streptomyces griseus*, subspecies *griseus* (ATCC 3343), purchased from the American Type Culture Collection (Manassas, VA). This specific culture is known to produce MIB and geosmin. The pipeloops were inoculated with actinomycetes prior to performing these experiments. The actinomycetes stock cultures were propagated using solution of Difco™ ISP Medium 1 (Sparks, MD). To inoculate the pipeloops 1 L of actinomycetes stock culture was centrifuged at 6500 rpm for 10 minutes. The supernatant was discarded and the actinomycetes pellet was resuspended in 405 mL of nanopure water and mixed well by stirring for five minutes. Prior to seeding actinomycetes, 5 mL of stock was taken out for titer determination. Finally, 200 mL of sample was added to each pipeloop.

Sampling ports were chosen in order to reflect different sections of a water distribution system. In order to maintain consistency and ensure the sample represents the flowing water each sampling port was flushed prior to sampling. The volume of flush water for each sampling port was determined based on its configuration. Table 2.1 exhibits the flushing volume for each sampling port.

Water samples were collected in sterile 50 mL conical tubes (VWR, West Chester, PA). MIB and geosmin samples were collected in 40 mL amber glass bottles. To preserve the samples 0.5 mL of 1 g/L sodium azide (Na_3N) was added to each sample after collection. Samples were stored at 4° C until they were assayed for MIB and geosmin using a Varian Star 3400 CX Gas Chromographer and a Saturn GC/MS/MS 4D Mass Spectrophoner (Sugarland, TX). The SPME fiber used was a Supelco #57348 U (Bellenfonte, PN).

Prior to seeding the pipeloop with an actinomycetes culture, background samples were collected and analyzed from ports A, B, and C. All samples were analyzed for MIB, and geosmin but only samples from ports A, B were analyzed for actinomycetes.

Biofilm samples were collected from both the cast-iron and PVC pipeloops. The sampling events in each system (PVC and cast-iron) rotated between four straight horizontal flow pipe segments. In order to sample the biofilm, the selected pipe segment was removed from the loop. To facilitate the removal of the pipe segment the system had to be shut down and partially drained. Overflow water was collected and stored in a ten-gallon bucket while the sampling was taking place. In this manner, the overflow could be put back into the system and thus, have the least amount of impact on the system.

The biofilm sampling protocol consisted of scraping a 25 mm deep segment around the entire inside circumference of both pipe ends with a piece of glass slide (25 mm by 75 mm). The biomass from both pipe ends was then combined resulting in a total area of 4160.5 mm^2. Each sample area was carefully washed with nanopure water subsequent to removing the biofilm. After combining both biofilm samples in sterile 50 mL centrifuge tubes, it was vortexed for ten minutes. This was done to disintegrate the biofilm and make a homogenous solution. Samples were serially (10-fold) diluted before analyses. One mL of each dilution was analyzed by double layer starch-casein media and plates were read after incubation at 28° C for seven days.

FIELD SAMPLE COLLECTION

Distribution System Site Descriptions

Twenty distribution samples were collected monthly for eighteen months to identify factors that produced T&O problems in the distribution system. Each sample was associated with one of the five City of Phoenix Water Treatment Plants. Four sites were sampled for each plant. Since it was assumed that time in the distribution system could be a factor of T&O, the sites were varied in distance from the plants. One site per plant represented a maximum retention site.

Watershed Sampling Site Descriptions

The City of Phoenix water supply comes from a combination of sources including the Verde and Salt Rivers and the Central Arizona Project (CAP) (Figure 2.3). Lake Pleasant, Bartlett and Saguaro Lakes are terminal reservoirs of this system. Water is released from the reservoirs and conveyed through open-channel, concrete lined canals. Canal walls provide miles of surface area for periphytic (attached) cyanobacteria to proliferate. During the warmer months, reservoirs thermally stratify approximately 10 m from the surface. As water temperatures decrease in the fall of each year, wind facilitates destratification (turn-over) bringing nutrients from below to the surface. Reservoirs are operated differently, and are subject to various upstream impoundments that may affect the growth of microcystin (MC)-producing cyanobacteria (Nguyen et al. 2002).

Lake Pleasant is an off-stream reservoir that stores a combination of CAP and Agua Fria river water. During the months of October through April, Lake Pleasant is filled with Colorado River water from the CAP canal. Water is released back into the canal between April and October. The Agua Fria River provided less than 5% of the total inflow to the reservoir during the three years of this study. Based on controlled water release, Lake Pleasant water levels and storage volume vary significantly (approximately 50%) throughout the year.

Bartlett Lake is an in-stream reservoir on the Verde River. The outlet is located near the bottom of the reservoir. Snowmelt at higher elevations is the source of the majority of the water entering the reservoir. Water is stored in Bartlett Lake during the summer months and released after the first of October and continued through the winter.

Saguaro Lake is located along the Salt River and is the lowest of 5 reservoirs. Saguaro reservoir is operated in series with an upstream reservoir to produce hydroelectricity during the high energy demanding summer. Water is released from Saguaro during the day, and then pumped back to the preceding reservoir during the night. This procedure creates a relatively short hydraulic retention time for the reservoir. Saguaro Lake is typically filled to capacity throughout the year. Most releases from this reservoir occur during the summer (May through September) and stop in early October. Table 2.2 provides a summary table of the physical characteristics of Bartlett, Saguaro, and Lake Pleasant during the 3-year span of the T&O research conducted at ASU.

The year round average air temperature in Phoenix is 23° C. Table 2.3 provides a summary of the average water temperatures near the surfaces of Lake Pleasant, Bartlett and Saguaro Lakes. The water temperature averages and ranges were calculated using data from the top 10 m of the reservoirs. Based on warm temperatures and clear skies, sunlight is readily available for photosynthetic process of cyanobacteria. The physical characteristics of the City of Phoenix water supply in part account for the large amount of cyanobacteria growth.

Table 2.1
Flush volumes by sampling port

Sampling Port	Flush Volume [mL]	Comments
Port#7	None	Used pipette to take sample from middle of reservoir
Port#1	1000	To flush stagnant water in pipe and the valve
Port#3	500	Flushed half of water from the configured pipe section

Table 2.2
Reservoir physical characteristics and hydrologic data at maximum pool elevations

	Bartlett Lake	Saguaro Lake	Lake Pleasant
Surface area (km^2)	11.2	5.1	40.3
Volume (m^3)	$2.2x10^8$	$0.85x10^8$	$10.5x10^8$
Depth (m)	38	33	43
Surface Elevation (m)	532	466	519
Hydrology during study			
Average Annual Inflow (m^3)	$2.8x10^8$	$3.7x10^8$	$7.2x10^8$
Average Annual Outflow (m^3)	$2.6x10^8$	$3.6x10^8$	$7.0x10^8$
Average HRT (days)	139	80	438

Table 2.3
Water temperatures near the surface of three reservoirs for the Phoenix metro area (Aug 1999- Jun 2002)

Reservoir	Water Temperature (° C)		
	High (August)	Low (January—February)	Average
Bartlett Lake	29	9	18.9
Lake Pleasant	29	11	19.4
Saguaro Lake	29	9	20.1

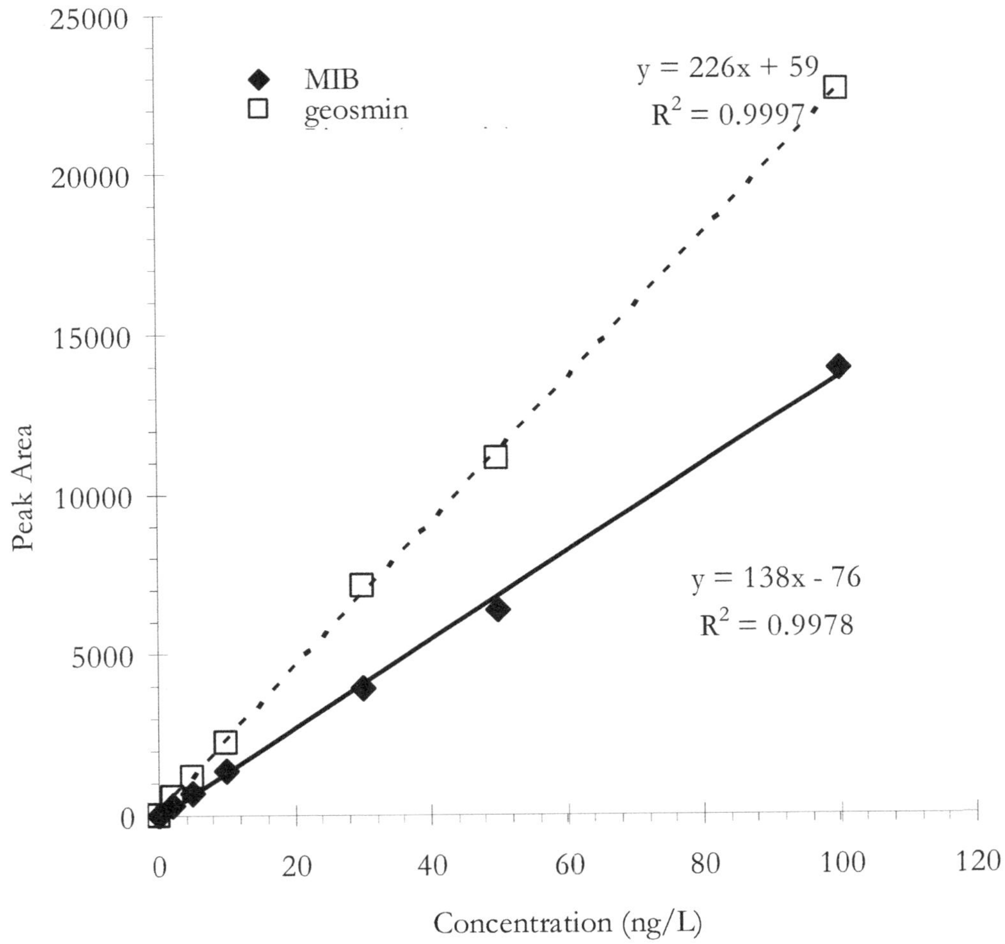

Figure 2.1 MIB and Geosmin standard calibration curve (08/13/02)

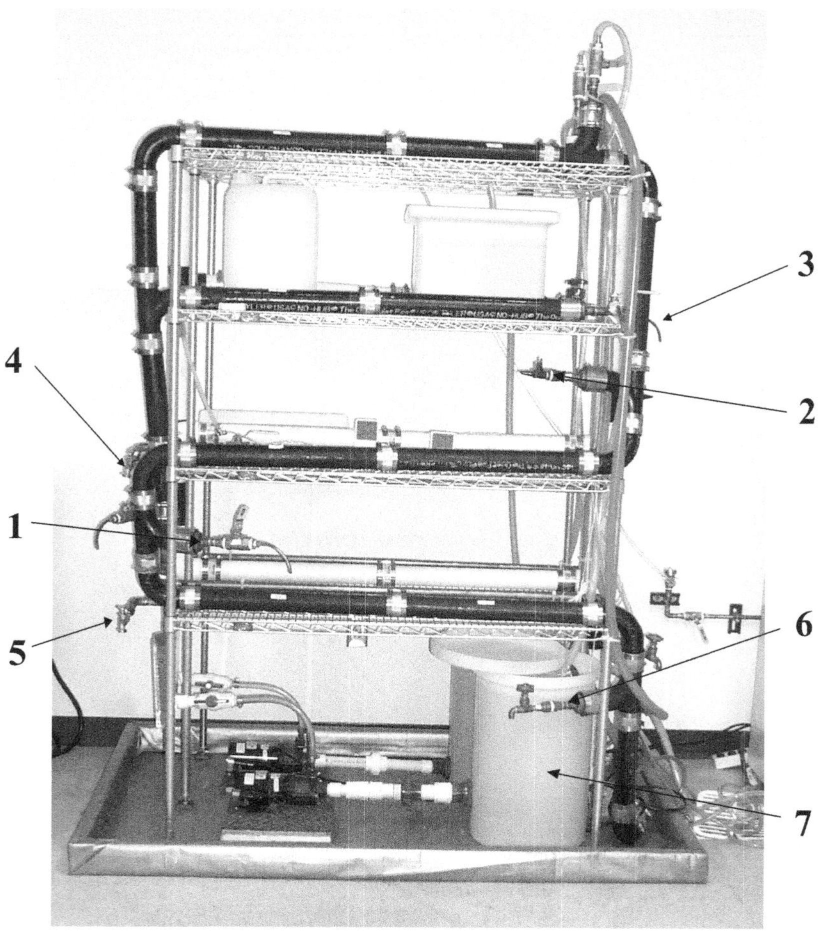

Sampling port	Description
1	Gate valve, unthreaded type, soldered.
2	PVC laboratory-style stop-cock valve
3	Dedicated sampling tap at the terminus of the dead end line
4	Gate valve (hose bib type), ¾ " diameter, copper service line, 6 " length
5	Gate valve (hose bib type), ¾ " diameter, copper service line, 18' length
6	Dedicated ball valve faucet sampling tap
7	Reservoir/ Influent

Figure 2.2 Schematic of laboratory pipeloop system and summary of system components.

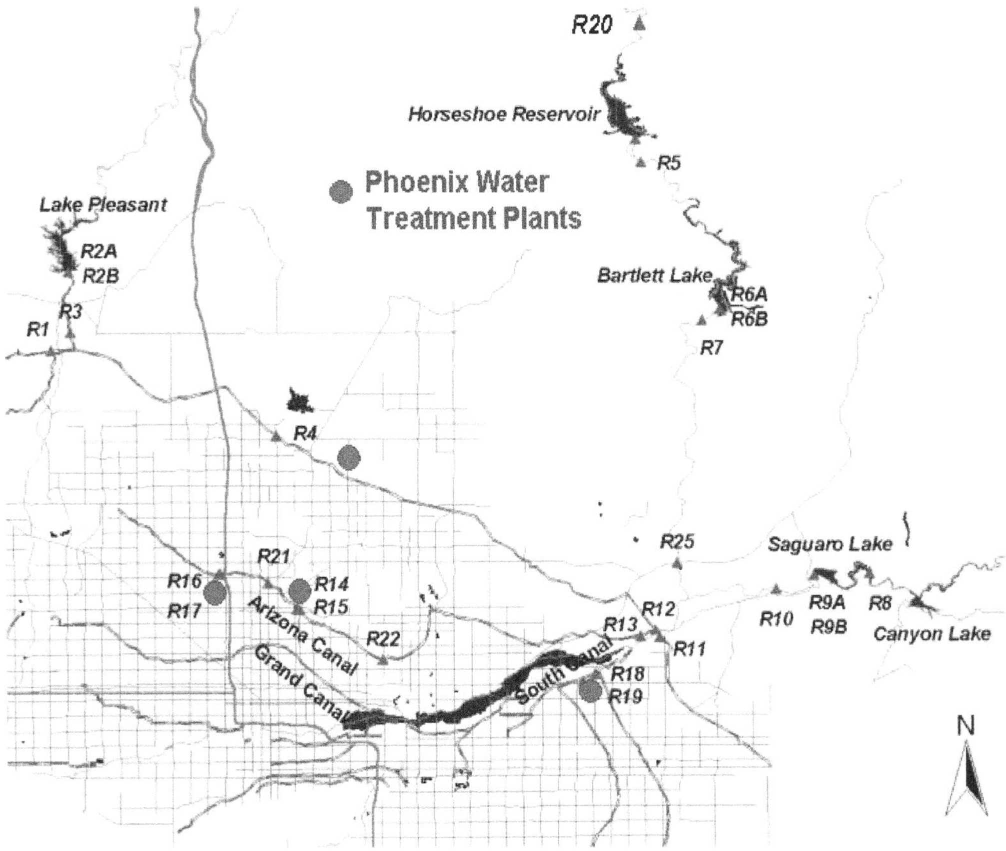

Figure 2.3 Monitoring sites for MC analysis (▲R#) and location of City Of Phoenix water treatment plants (●). Samples analyzed for MC concentration included reservoir samples R2, R6, and R9; canal samples R11, R12 and R13; and Deer Valley and 24th Street treatment plants

CHAPTER 3
MICROCYSTIN METHOD VALIDATION AND COMPARISON

This chapter provides a brief background on Microcystin analysis and then details methods and data validation for an ELISA based Microcysin method which was considered the easiest to apply in WTP labs. In addition to processing commercial standards, an algae culture known to produce Microcystin was obtained and grown to investigate intracellular and extracellular Microcystin.

BACKGROUND

Several methods are available for monitoring Microcystin (MC) in the water supply. These techniques include: high performance liquid chromatography (HPLC), microcystin enzyme linked immunosorbent assay (ELISA), and protein phosphatase-1 or -2A (PP1 or PP2A) inhibition assay.

HPLC is an effective method in determining MC concentration. This method has been used as a standard in development of detection techniques. The principle behind HPLC is separation of toxins from other co-extracted compounds based on compound molecular weight. HPLC is typically followed by ultraviolet (UV) absorbance or mass spectrometry (MS) to identify MC. These methods are highly sensitive. Nicholson et al. 2001 were able to achieve a detection limit of 0.02 μg/L using HPLC-MS. Extraction of the cyanobacterial mass prior to HPLC may create quantification and MC recovery problems (Nicholson et al. 2001). Fischer et al. 2001 summarized several of the limitations of the HPLC methods. Natural organic matter (NOM) may mask toxins in HPLC testing. Also, HPLC is capable of detecting single variants of microcystin, but quantification of all present congeners is difficult. Individual MC variants may be present at concentrations below chromatographic detection limits, but the possibly significant combined toxicity of these variants would not be measured (Fischer et al. 2001). The biochemical assays, ELISA, PP1 and PP2A, are less qualitative than HPLC, but are as sensitive, account for combined toxicity, and provide more rapid results (Carmichael et al.1999). For these reasons HPLC was not examined as a monitoring method for the City of Phoenix and will not be employed in this study.

The commercially available ELISA kits are a second MC detection method. These kits operate based on the competition of MC in the sample with an enzyme-labeled MC for a limited number of antibody binding sites. The concentration of MC is then determined colorimetrically. In Step #1 an assay well is coated with an antibody specific to MC. Step #2 involves the injection of MC standards and samples to the well. Microcystin present in the standards and samples binds to the antibodies. An enzyme-linked antigen is added in Step #3 and binds to any unoccupied antibody binding sites. In Step #4 an enzyme substrate is added to the well. The substrate binds with the antigen and produces a color. The amount of color present indicates the amount of MC present in the standard or sample. The darker the color of the well, the less MC is present. The cross reactivity of ELISA varies from good (MC-RR) to poor (MC-LA) for the MC congeners. ELISA tests may underestimate MC concentration due to poor cross-reactivity among variants (Nicholson et al. 2001). Cross-reactivity is a function of structural similarity that does not necessarily represent similar toxicity (Nicholson et al. 2001). Thus ELISA may be best employed as a semi-quantitative assay, or a screening tool. Rivasseau et al. 1999 found organic interferences that lead to false positives in MC spiked water. Based on these interferences, the limit of quantification for the assay was 0.2 µg/L. EnviroLogix Inc.™7, the manufacturer of the kits employed in the ASU research, lists a limit of detection for their MC ELISA kits at 0.147 µg/L. These kits were used at ASU to measure MC concentrations in the water supply reservoirs, and as a standard to refine the methods used in the PP2A assay.

The third method for determining MC concentration is the protein phosphatase assay (PP1 and PP2A). Nicholson and Shaw, 2001 explained the principles behind these assays. The hepatotoxins, MC and nodularins, inhibit enzymes that control the dephosphorylation of intracellular phosphoproteins. These enzymes are Type 1 (PP1) and Type 2, including a subset of Type 2 called protein phosphatase-2A (PP2A). Based on this process, the inhibition of these enzymes by MC and nodularins provides a measure of toxin concentration. The assay measures the phosphate release from a phosphorylated protein in the presence of PP1 or PP2A enzyme and an inhibitor (MC) (Nichloson et al. 2001). Complications arose in the measurement of the phosphate released due to the short half-life of the ^{32}P isotope. In order to overcome this challenge, the substrate p-nitrophenyl phosphate was used in the assay, and the color released from the substrate in the presence of MC and PP2A was measured (An et al. 1994). The sensitivity of this assay is in the part per billion range (An et al. 1994). Heresztyn and Nicholson, 2001 reported a working range for the assay of 0.05 - 1.0 µg/L. The PP2A enzyme is preferred over PP1 because it has been reported to be 50 times more sensitive in inhibition of MC (Honkanen et al. 1990; Honkanen et al. 1994). PP2A is being employed and adapted by researchers at ASU to develop a detection method that incorporates current lab equipment and techniques used by the City of Phoenix water utilities.

Solid phase extraction with C18 cartridges provide a means of improving the MC detection capabilities. ELISA and PP2A are able to detect MC in the part per billion range. Carmichael and An, 1999 provided three reasons for using C18 cartridges in toxin recovery. They were used to:

1. Lower pigments in sample extracts
2. Remove inorganic compounds that also act as protein phosphate inhibitors and would result in false positives for MC
3. Concentrate MC and nodularin toxins in samples.

Several aspects of the ELISA and PP2A assays were considered for analysis including:

- Detection range and method detection limits (MDL)
- Matrix interference
- Freezing effects
- Assay consistency
- C18 cartridge recovery capabilities
- Intracellular extraction techniques

Microcystin Analysis

Two known MC producing cyanobacteria species were selected for use in the development of intracellular MC extraction techniques. Two *Microcystis aeruginosa* cultures (B2666 and B2667) were purchased from the University of Texas at Austin. Algal cells from the UTEX cultures were transferred into autoclaved, glass 250 mL flasks containing approximately

150 mL of media J (Table 3.1). Foam stoppers were used to prevent culture contamination and allow air passage. Cultures were placed on a platform shaker table (Lab-Line Instruments, Inc) and continually mixed at a speed of 100 rpms under a fluorescent light providing approximately 20 $\mu E/m^2/s$. As the cultures matured, approximately 2 mL of the culture were transferred to a new 250 mL flask of media J to keep cultures viable.

Intracellular Extraction

Three methods of intracellular extraction were compared to determine the procedures for obtaining maximum MC extraction (Figure 3.1). Two milliliters of *Microcystin aeruginosa* culture were passed through a syringe filter (ashed GF/C Whatman). The eluent was set aside for extracellular MC analysis with ELISA.

Method A consisted of extraction solely through beadbeating (mini-beadbeater 3110BX BioSpec Products, Inc). The biomass captured on the filter paper was washed into a beadbeater tube. Silicon beads of 0.1 mm diameter were used in the beadbeating process. Methanol (MeOH) was added to the beadbeating vial to ensure there was no headspace remaining. The MeOH MC mixture was decanted and diluted with Super Q™ to create solution containing less than 20% MeOH to avoid MeOH interference with ELISA. Method B measured the extraction of intracellular toxins through cellular membrane disruption by MeOH followed by beadbeating. For both methods B and C, the GF/C filter was soaked in MeOH and Super Q™ laboratory water for 24 hours. In B 1.5 mL of the mixture was removed and passed through the same beadbeating and dilution process as discussed in A. Method C was cellular contact with MeOH for 24 hours. For method C the remaining 10 mL of MeOH/Super Q™ were centrifuged to remove cellular and filter debris from the solution. The supernatant was decanted then blown down to 2 mL with nitrogen. The solution was then diluted with Super Q™ water. For samples containing sufficient color, C18 extraction may be used (Figure 3.2). ELISA was used to analyze the MC concentrations in each extract.

Solid Phase Extraction

General MC recovery capabilities of C18 cartridges were examined using extractions of Super Q™ water spiked with microcystin-LR (L = 1-leucine, R = 1-arginine) run in triplicate. A standard of 1.2 µg-MC/L was prepared for C18 concentration using MC-LR stock (Calbiochem-Novabiochem, Corporation). Six mL of the standard were set aside, then the remaining 3 L were passed through three DSC-18 cartridges (Supelco, Supelco Park, Bellefonte, PA, U.S.A. 505 m^2/g specific surface area, 57 µm average particle diameter, 67 angstroms average pore diameter). The C18 concentration process is illustrated in Figure 3.2. Step #1 illustrates the conditioning of the C18 cartridges by gravity feeding MeOH and Super Q™ water through the cartridge. In Step #2 one liter of the standard was passed through each C18 cartridge. The cartridge was then rinsed in Step #3 to remove any color introduced by natural organic matter. The cartridges were eluted in Step #4 using 10 mL of 80% methanol solution and collected in 40 mL vials. The solution was then blown down to 2 mL in Step #5. Finally, in Step #6, the MC concentration in the reservoir samples and the concentrated standard solution were determined with PP2A and ELISA assays. Lake samples passed through C18 cartridges were filtered (Whatman GF/F) then concentrated using the same C18 procedures described above.

Enzyme-linked Immunosorbent Assay (ELISA)

Recommended methods for the EnviroLogix™ ELISA kit were followed for determining MC concentration in various samples. The ELISA kit consisted of an assay plate of 96 antibody coated wells, MC standards of 0.16, 0.6, and 2.5 μg MC/L, a negative control, assay diluent, MC-enzyme conjugate, wash solution, substrate, and stop solution. Initially, the plate, reagents and standards were allowed to reach room temperature. Using a multi-channel pipette, 125 μL of MC assay diluent were added to each of the wells, followed by 20 μL of the standards, negative control, and samples in their respective wells. Each of the controls, standards, and samples were tested in triplicate to account for result variations and experimental errors. The plate was then covered with Parafilm to prevent evaporation and placed on a platform shaker (New Brunswick Scientific) for 30 minutes. After mixing, 100 μL of MC-enzyme conjugate were added to each well. The plate was then placed on the shaker and stirred for another 30 minutes. After the second mixing period, the Parafilm was removed and the well contents were shook into a sink. The wells were next flooded with the wash solution then emptied by inverting the plate and shaking the contents into the sink. This flooding and shaking procedure was repeated four times. Removal of as much wash solution as possible was accomplished by slapping the plate on a paper towel. Then 100 μL of substrate were added to each well. The plate was again covered and placed on the shaker for 30 minutes. Finally, 100 μL of stop solution were added to each of the wells. The optical density (OD) of the wells was measured using a microplate reader (Molecular Devices kinetic microplate reader). The wavelength of the plate reader was set at 450 nanometers with a reference wavelength of 650 nanometers. A column of wells was filled with 200 μL of DI water to provide a zero for the microplate reader. Microcystin concentrations were determined by developing a semi-log curve of $\%B_o$ (Equation 3.1) versus MC concentration of the standards (Figure 3.3). The ELISA standards were chosen to target the linear range of this curve. The equation of best-fit line for the linear range is then used to calculate MC concentrations in the samples. For the assay data used in developing Figure 3.3, the R^2 value of 0.9976 indicates a tight correlation between the trendline and the $\%B_o$ values.

$$\%B_o = \frac{B}{B_o} \times 100\% \qquad \text{Equation 3.1}$$

Where: B = optical density of sample or standard
B_o = average optical density of negative control

Protein Phosphatase –2A Assay (PP2A)

The procedures outlined by Carmichael and An, 1999 for PP1 analysis were combined with results using PP2A enzyme conducted by Heresztyn and Nicholson, 2001 to develop a PP2A assay. This method involves the preparation of a diluent buffer, enzyme reagent, and substrate. The diluent buffer was prepared using 4 mL of 1 M Tris-HCL (pH=8.6), 2 mL of 1 M KCl, and 3 mL of 1 M $MgCl_2$. Intermediate solutions A, B, and C were prepared by adding 4.6 mg Dithiothreitol (DTT), 15 mg bovine albumin (BSA), and 15 μL of 1 M $MnCl_2$ to 10 mL of diluent buffer respectively. Solutions A, B, and C were then combined creating a total reagent volume of 30 mL. The enzyme reagent was then prepared by spiking 5.3 μL of PP2A stock enzyme (Promega Ppase-2A) into 4 mL of reagent. Next 63 mg of phophatase substrate (Sigma 104 Phosphatase) was added to 6 mL of reagent creating the substrate solution. A 96-well plate (Corning Inc Costar, cell culture cluster, flat bottom Tissue Culture treated) was used for the PP2A assays. Next 40 µL of enzyme reagent were added to each well, followed by 10 µL of MC standards, DI water, and unknowns in triplicate in their respective wells. The plate was covered with Parafilm and placed on the platform shaker for 20 minutes. After 20 minutes 50 µL of substrate were added to each well. Bubbles in the wells were popped using a wire, then 100 µL of DI water were added to empty wells to be read as blanks by the microplate reader, and the plate was placed in the microplate reader. The absorbance at 405 nanometers for each of the wells was measured at 5-minute intervals for 1 hour. The maximum optical density reading changes over time, V (mOD/min), were recorded. The ratio of sample color change (Vi) over color change in the zero MC wells (Vo) was calculated. MC concentration versus the ratio of Vi/Vo was then plotted to develop a standard curve.

ASSAY DETECTION RANGE AND LIMITS

The EnviroLogix™ ELISA kit specifies a MDL of 0.147 µg/L for MC. The MDL was calculated using an interpolation at three standard deviations from the mean (81.3% B_o) for negative water samples. The MDL values calculated using the procedures outlined in the Code of Federal Regulations (1986) were 0.21 µg-MC/L (3 times the standard deviation for eight replicate samples), and 0.24 µg-MC/L (based on 95% confidence using the F-ratio method). The kit MDL of 0.147 µg-MC/L was used in reporting sample MC concentrations. Standard curves developed for ELISA ranged from 0.16 to 1.6 µg-MC/L in kits used from September 2001 through August 2002. In succeeding months EnviroLogix™ increased the range of the standard curve from 0.16 to 2.5 µg-MC/L.

The ELISA assay provided a consistent, reliable method for monitoring MC concentration in the reservoirs. Table 3.2 summarizes the ELISA assays run over the 14 months. EnviroLogix's™ kits provide removable rows of wells with each tray. When an assay is run, the user determines the number of rows needed for the assay, therefore more than one assay may be performed using the same kit. The column Kit No. in Table 3.2 indicates the kit number assigned by ASU to track any trends specific to an individual kit. For example, in the initial assays (#1-4), the 50% B_o was less than 0.6 µg-MC/L, the 50% B_o value determined in assays #5 -11. Kits (#1 and #2) were near the manufacture's specified expiration date, and this variation may be due to some degradation of the substrates, enzymes and antibodies. The ELISA assays provided consistent readings as seen in the similar slopes and high R^2 values. All of the ELISA assays were considered acceptable based on high R^2 values (all greater than 0.94), and consistent % Bo standard values. The Envirologix™ kit is a reliable method for determining MC concentration.

The MDL and range of the PP2A assay are dependent on protein phosphatase-2A enzyme reactivity. Initially the MDL for PP2A was 0.05 µg-MC/L calculated according to procedures outlined in the Code of Federal Regulations (1986), using three sample concentrations of 0.15, 0.3, and 0.02 µg-MC/L. The initial MDL tests were conducted using a PP2A assay range from 0.05 to 0.25 µg-MC/L and are valid for assays within this range. However, enzyme reactivity was inconsistent from assay to assay. As a result PP2A assay ranges varied with each assay performed. A more typical range was from 0.24 to 2.5 µg-MC/L. The MDL for assays in this range was 0.24 µg-MC/L.

Enzyme stability proved to be a hindrance with PP2A analysis. Table 3.3 summarizes the PP2A assays run over a period of seven months. The Enzyme No. column indicates the ASU assigned number to separate assays run with the same enzyme. Protein phosphatase-2A is stable at –20° C. As the enzyme warms, it activates and should promptly be used in analysis. The enzyme stock used in the lab (Promega PPTase-2A, V631A) was difficult to keep below this temperature. The enzyme variations resulted in unpredictable 50% Vi/Vo values, and various assay ranges. The MC concentrations of several samples were undetermined due to the inconsistent assay ranges. Assays were considered unacceptable based on inconsistent calibration results, lack of color production, or poor R^2 values (< 0.75). These concentrations may have been detectable with ELISA, or with different protein phosphatase-2A enzyme. Figures 3.4 - 3.6 are calibration curves developed using three different PP2A enzymes. Figure 3.4 was developed using enzyme #1 that was initially warmed to room temperature before use in the assay. The high R^2 value indicates consistency between calibrations. The MDL for this assay was 0.05 µg-MC/L. Figure 3.5 was developed using enzyme # 4. Again, a high R^2 value is observed. The MDL for this assay was approximately 0.24 µg-MC/L, and the assay range is much larger than in Figure 3.4. Enzyme # 4 was kept at –20° C and was not allowed to warm, resulting in greater enzyme activity and thus the increased assay range. Figure 3.6 was developed using a fresh enzyme (# 5) that was also stored at –20° C. The R^2 value is lower than expected and there is a usual amount of scatter between calibration replicates. The inconsistency of PP2A enzyme proved to be a problem. It was difficult to select appropriate calibration concentrations when such large variations in assay ranges occur and are difficult to anticipate. Therefore, the ELISA assay is recommended for future monitoring based on its consistent assay results.

MATRIX EFFECTS

The potential effects of the City of Phoenix water matrix on ELISA and PP2A assays were examined. Three samples from each lake were filtered (Whatman GF/F) and spiked with MC stock solution to obtain MC concentrations of 0, 0.5 and 1.0 µg-MC/L. Control samples were prepared similarly in Super Q™ water spiked with the same MC concentrations. Samples were analyzed using ELISA and PP2A to determine potential matrix influences. No distinct trends were identified for a particular water matrix (Table 3.4). The MC concentrations presented in Table 3.4 are the average of three assay wells. No trend is noted between ELISA and PP2A results. That is, ELISA results are neither consistently greater than nor less than the PP2A results. The Super Q™ sample for the 0.5 µg-MC/L dose was undetectable according to ELISA and at a lower concentration than similarly spiked lake samples. There was no apparent reason for this low value. The amount of deviation between the different samples of the same dose illustrates the variability within the assays. Based on these results, matrix effects of natural waters in Arizona on the PP2A and ELISA assays appear to be negligible.

FREEZING EFFECTS

Collected water samples were frozen and stored at –20° C. Cell lyses may occur due to freezing of the water samples (Yoo et al. 1995). The largest MC toxin fraction is contained within cyanobacterial cells (Yoo et al. 1995). Thus cell lysis encouraged through freezing should result in increased MC concentrations. During the increased MC production period observed from May through October, extracellular MC concentrations measured initially were not consistent with the concentrations measured after freezing. In all cases the concentrations decreased. Based on the inconsistent results between unfrozen and frozen samples, an experiment was conducted to determine possible freezing effects on MC concentration in natural water. The tests were performed using MC stock solution. Any cell lyses occurring would be accounted for within the specific lake water matrix. One-liter samples from each lake and Super Q™ were spiked with MC stock to obtain MC concentrations of approximately 1.0 μg-MC/L. Samples were analyzed using PP2A, frozen and stored in a freezer for two weeks, then analyzed using ELISA. Results are presented in Table 3.5. Based on a Student t-test with an α-value of 0.05, there is a freezing effect on MC concentration in the samples. For the individual matrices, Pleasant and Saguaro samples had the greatest difference between frozen and unfrozen MC concentrations. In nine of the twelve samples, the frozen samples resulted in lower concentrations than the unfrozen samples. This phenomenon has not been documented in the literature and may be attributed to differences in the specific sample matrices, enzyme activity, or PP2A and ELISA producing different concentrations. Further research is required to determine what characteristics of the matrices are causing the decrease in MC concentration when the samples have been frozen. Variability in ELISA and PP2A assays are discussed in the next section of this report.

ASSAY COMPARISONS

The calibration standards from an ELISA kit were tested to determine the accuracy of PP2A analysis in detecting MC concentration. The ELISA standard concentrations were 0.16, 0.5, and 1.6 μg-MC/L. The 0.5 and 1.6 μg-MC/L standards were diluted 10:1 to bring the MC within the detection limits of PP2A (0.05 to 0.25 μg-MC/L for this enzyme). The standards were treated as unknown samples and a PP2A assay was used to determine MC concentration. The ELISA standards were tested on this PP2A plate to compare the MC concentration measurements of the two tests (Table 3.6). The PP2A test reasonably predicted the MC concentration of the standards with the tightest correlation for the 0.16 μg-MC/L standard and the greatest variability for the 0.5 μg-MC/L calibrator. The 0.16 μg-MC/L standard was already within the detection limits of the PP2A assay and therefore no dilution was necessary. Dilution of the 0.5 and 1.6 μg-MC/L standards may account for the greater variation among the replicates. Figure 3.7 provides a comparison of the two MC concentrations. Perfect agreement between the tests would have resulted in a curve with a slope and R^2 value of one. In this case the R^2 value is 0.88, indicating some deviation between the tests. In this case the slope was 0.87 indicating the PP2A measured concentrations were less than specified ELISA concentrations. The percent error was calculated by subtracting the PP2A concentration from the ELISA specified concentration and dividing the difference by the ELISA concentration. The percent error column illustrates the larger deviations for the diluted 0.5 and 1.6 μg-MC/L standards. Overall the PP2A assay reasonably predicts MC concentration in undiluted samples.

Several samples were analyzed using both the PP2A and ELISA assays. These samples are compared in Figure 3.8. Slightly greater variability is seen in the PP2A analysis. The 1:1 sloped line provides an indication of perfect correlation between the assay results. The majority of the data falls near this line, indicating reasonable correlation between the MC concentrations determined by the assays.

SOLID PHASE EXTRACTION

The recovery capabilities of C18 extraction were tested using prepared 1.2 μg - MC/L samples. The detectable range for this PP2A assay was 0.07 - 0.26 μg/L. Dilutions of both the initial sample and C18 concentrated standards were used in PP2A analysis. Reservoir and Deer Valley samples were not diluted based on previous undetected levels of MC in these waters. Table 3.7 presents the results of C18 extraction of spiked Super Q™ water. The results of the C18 extraction were somewhat varied, recovering 1.2 ± 0.42 μg-MC/L. This variation in recovery capabilities may be attributed to different cartridges and should be considered when examining MC concentrations obtained from extracted samples. Standard # 1 showed 68% recovery; Standard # 2 showed 92% recovery, and Standard # 3 indicates 133% recovery. Based on the variability seen in these results, future C18 concentrated results should be viewed as approximate concentrations, not exact values. The MC concentration examined utilizing C18 recovery will typically be in the part per trillion range. This is a thousand times lower than the World Health Organization (WHO) limit. The variable recovery among the C18 replicates is not a concern because the approximate values are significantly less than toxic concentrations that pose health risks and are still useful for MC monitoring.

Four samples from the City of Phoenix water supply were concentrated using C18 cartridges and tested for MC using the PP2A assay. The MC concentration in the samples was anticipated to be low based on previous screening work, therefore no dilution of the C18 concentrates were performed. Table 3.8 illustrates the increased detection capabilities C18 concentration provides for MC monitoring of the City of Phoenix water supply. During the month of May, Bartlett Lake, Lake Pleasant, and Saguaro Lake have MC concentrations below the detection limits of both the PP2A and the ELISA tests. By using the C18 cartridges to concentrate the MC in the reservoir samples, the amount of MC in the Bartlett, Lake Pleasant, and Saguaro reservoirs was determined to be 2.1, 2.9 and 2.4 ng-MC/L respectively. The Deer Valley sample during the month of April had a concentration of 3.2 ng-MC/L. Thus through the use of C18 cartridges, MC present at low concentrations are detectable. The results indicate the presence of MC toxin in Bartlett, Lake Pleasant, and Saguaro reservoirs as well as in the canals at part per trillion levels. Thus C18 concentration allows the monitoring of fluctuations in MC reservoir at levels less than the MDLs of ELISA and PP2A assays.

INTRACELLULAR EXTRACTION

Three procedures were tested to determine the most complete MC intracellular extraction technique. The three methods included (A) direct beadbeating of filtered biomass, (B) 24-hours in methanol (MeOH) followed by beadbeating of filtered biomass, (C) and 24-hours in MeOH. The extractions were performed on a culture of *Microcystis aeruginosa* (UTEX) after 15 days of growth. The extracellular concentration of the sample was 0.62 μg-MC/L determined by ELISA, and the intracellular concentration 494 ± 89 μg-MC/L. An additional test was run to further compare method (B) and method (C).

Intracellular extraction tests indicated that cells soaked for 24 hours in MeOH followed by cell lyses through beadbeating (Method B) provided the most robust extraction. In the initial intracellular extraction comparison, Method C, MeOH soaking followed by C18 extraction, measured the highest intracellular concentration of 0.795 μg-MC/OD_{730}. Method B measured a concentration of 0.627 μg-MC/L, and Method A, direct beadbeating, a concentration of 0.533 μg-MC/OD_{730}. A second extraction test was done to further compare Methods B and C.

The second test was conducted on the same *Microcystis aeruginosa* culture this time after 10 and 29 days of growth. Table 3.9 and Table 3.10 present the results for the intracellular extraction methods. The variation in the intracellular concentrations may be explained, in part, by the heterogeneous nature of the culture. Cells tended to form "clumps" when the culture was disturbed. Thus it is likely that different amounts of biomass were removed during the initial filtration of these samples. Direct beadbeating (A) removed significantly less intracellular MC than both methods B and C. Method B provided the most robust extraction of intracellular MC, however, method C produced similar results and is the simplest extraction technique. Based on this analysis, method C is recommended for use in future intracellular extraction. These concentrations were calculated using Equation 3.2.

$$C_{OD} = \frac{C_{ELISA} \text{ x } V}{OD_{730}} \qquad \text{Equation 3.2}$$

Where: C_{OD} = MC concentration of sample (μg-MC/OD_{730})
C_{ELISA} = concentration determined through ELISA assay (μg-MC/L)
V = filtered culture volume (L)

In the second extraction, Method B measured the highest intracellular concentration. However, Method C also provided similar intracellular extraction, and is recommended for future extraction based on method simplicity and sufficient extraction.

The amount of biomass present in a sample may be approximated using the OD_{730} of the sample. For the *Microcystis aeruginosa* samples, a relationship was developed between the optical density of a culture and its corresponding biomass. Dry biomass was determined for OD_{730} values for *Microcystis aeruginosa* culture after 21 and 41 days of growth (Table 3.11). The dry mass per liter of culture was also determined. Figure 3.9 compares biomass concentration with OD_{730}. As anticipated, biomass concentration increases with increasing OD_{730} values. Using a best-fit curve for this figure, biomass concentration may be estimated from the OD_{730} using equation 3.3.

$$OD_{730} = 1.694\,(B) + 0.456 \qquad \text{Equation (3.3)}$$

Where: OD_{730} = Optical density of sample measured at a wavelength of 730 nm (1/cm)
B = biomass concentration per liter of solution (mg-biomass/L)

Optical density and intracellular MC concentrations changed over time in algal cultures. The MC producing cultures indicated that the ratio of intracellular MC and OD was very similar when measured over time (Figure 3.10). Figure 3.10 also provides the growth curve of the culture with a lag phase up to day 10, followed by a period of exponential growth. Additionally, the MC concentrations measured using the three extraction methods are indicated in the figure.

SUMMARY

Assays and intracellular extraction methods were examined for future use in monitoring MC in the City of Phoenix water supply. The results of these analyses include:

- The ELISA assay has a MDL of 0.147 μg-MC/L, a range from 0.147 to 2.5 μg-MC/L. Assay results were unaffected by water matrices, and 100% of the ELISA assays conducted were considered acceptable based on their high R^2 values. ELISA assays are recommended for future MC analyses.
- The PP2A assay has a MDL 0.05 μg-MC/L for an enzyme range from 0.05 to 0.25 μg-MC/L, and a MDL of 0.24 μg-MC/L for an enzyme range from 0.24 to 2.5 μg-MC/L. The PP2A assay was unaffected by water matrices, and 74% of the assays conducted were considered acceptable. Assays were considered unacceptable based on inconsistent calibration results, lack of color production, or poor R^2 values (< 0.75). Enzyme activation at room temperature was considered the main source of error in PP2A assays. PP2A is not recommended for future MC monitoring.
- Freezing of samples before MC analysis affects measured MC concentrations, and further research should be conducted to determine the source of this concentration change.
- ELISA and PP2A predicted similar concentrations for three ELISA kit standards.
- SPE with C18 cartridges provides a means of measuring MC in the part per trillion range.
- Method C (24 hours in MeOH) is recommended for future intracellular extraction based on relative ease and completeness of extraction.

Table 3.1
Media J composition

Modification of Parker's J (or, Jansen) medium for *Microcystis*.
Preparation: Stock **Solutions C, D, and G9 trace metals** are prepared separately in 1000 mL quantities as described below and stored at 4° C. Combine 20 mL of **Solution C** with 980 mL of glass-distilled water, adjust to pH 9.0 with 1N NaOH and autoclave. Combine 20 mL of stock **Solution D** with 980 mL of glass-distilled water (do not adjust pH) and autoclave. After cooling, combine the two solutions 1:1 to make J medium.

Solution C
Combine the following compounds in the amounts shown below with glass-distilled water to make 1000 mL of solution C:

Compound	
$NaNO_3$	67.0 g
Na_2HPO_4	0.67 g
K_2HPO_4	3.8 g
Tricine buffer	25.0 g

Solution D
Combine the following compounds in the amounts shown below with 100 mL of G9 trace metals and glass-distilled water to make 1000 mL of solution D:

Compound	
$MgSO_4 \cdot 7H_2O$	10.0 g
$MgCl_2 \cdot 6H_2O$	5.0 g
$FeCl_3 \cdot 6H_2O$	0.5 g
$CaCl_2 \cdot 2H_2O$	4.2 g
Na_2EDTA	4.4 g

G9 trace metals
Combine the following compounds in the amounts shown below with glass-distilled water to make 1000 mL of G9 trace metals solution:

Compound	
H_3BO_3	3.25 g
$MnSO_4 \cdot H_2O$	1.5 g
$ZnSO_4 \cdot 7H_2O$	0.3 g
$(NH_4)6Mo_7O_{24} \cdot 4H_2O$	0.08 g
$CuSO_4 \cdot 5H_2O$	0.05 g
$Co(NO_3)_2 \cdot 6H_2O$	0.15 g
KI	0.01 g

Optional ingredients: In general, Microcystis grows better in liquid than on solid medium and does not respond well to commercial agar. Use electrophoresis grade agarose at 3.5 g/L, for growth on solid medium.

Table 3.2
Summary of ELISA assays conducted

Assay No.	*Date*	*Kit No.*	*50%Bo (μg/L)*	*Slope*	R^2	*Acceptable*	*Assay Range (μg/L)*	*Comments*
1	*11/27/02*	*1*	*0.4*	*-28.8*	*0.957*	*Yes*	*0.16-1.6*	*Calibrations less than recommended %Bo ranges. Near expiration date*
2	*1/22/02*	*1*	*0.2*	*-22.3*	*0.947*	*Yes*	*0.16-1.6*	
3	*2/02/02*	*2*	*0.2*	*-22.7*	*0.961*	*Yes*	*0.16-1.6*	
4	*5/15/02*	*2*	*0.2*	*-24.3*	*0.947*	*Yes*	*0.16-1.6*	
5	*6/25/02*	*2*	*0.5*	*-28.3*	*0.982*	*Yes*	*0.16-1.6*	
6	*8/28/02*	*3*	*0.6*	*-27.0*	*0.993*	*Yes*	*0.16-1.6*	
7	*9/12/02*	*3*	*0.6*	*-24.1*	*0.975*	*Yes*	*0.16-2.5*	
8	*9/19/02*	*4*	*0.6*	*-22.6*	*0.989*	*Yes*	*0.16-2.5*	
9	*10/10/02*	*5*	*0.6*	*-27.6*	*0.992*	*Yes*	*0.16-2.5*	
10	*10/17/02*	*6*	*0.6*	*-26.7*	*0.997*	*Yes*	*0.16-2.5*	
11	*10/25/02*	*7*	*0.6*	*-26.8*	*0.987*	*Yes*	*0.16-2.5*	

Table 3.3
Summary of PP2A assays conducted

Assay No.	Date	Enzyme No.	V/Vo= 50% µg/L	Slope	R^2	Acceptable	Assay Range (µg/L)	Comments
1	4/26/02	1	0.15	-3.12	0.9988	Yes	0.05-0.30	Enzyme at room temperature
2	6/06/02	1	0.15	-3.36	0.9415	Yes	0.05-0.25	
3	6/13/02	1	0.15	-3.03	0.9143	Yes	0.05-0.30	
4	6/20/02	2	0.15	-2.93	0.7110	Yes	0.05-0.25	Poor R^2 value
5	6/27/02	2	NA	NA	0.2489	No	NA	New enzyme used, calibrations not in range of Vi/Vo < 1
6	6/27/02	2	NA	NA	0.3110	No	NA	
7	7/10/02	2	NA	NA	NA	No	NA	
8	7/12/02	2	0.6	-0.54	0.9276	Yes	0.2-1.2	
9	7/12/02	3	0.8	-0.55	0.9741	Yes	0.2-1.2	
10	7/24/02	3	0.6	-0.52	0.9656	Yes	0.2-1.2	
11	7/24/02	3	0.7	-0.53	0.8232	Yes	0.2-1.5	
12	7/24/02	3	0.7	-0.52	0.9626	Yes	0.2-1.5	
13	7/24/02	3	0.9	-0.51	0.9306	Yes	0.2-1.5	
14	8/02/02	3	NA	NA	0.0085	No	NA	Enzyme was inactive from overuse at room temperature
15	8/02/02	4	1.1	-0.51	0.9027	Yes	0.2-1.6	
16	8/08/02	4	0.6	-1.03	0.9297	Yes	0.2-0.9	
17	8/21/02	4	1.4	-0.43	0.9588	Yes	0.3-1.8	
18	8/28/02	4	1.0	-0.47	0.9917	Yes	0.2-1.5	
19	9/12/02	4	1.0	-0.53	0.9310	Yes	0.4-1.5	
20	9/19/02	5	1.5	-0.19	0.7516	No	0.5-2.5	Inconsistent calibrator results
21	9/25/02	5	NA	NA	NA	No	NA	No color produced
22	9/26/02	4	1.2	-0.36	0.9416	Yes	0.5-2.1	
23	10/03/02	4	1.4	-0.35	0.9077	Yes	0.5-2.5	

Table 3.4
ELISA and PP2A results (µg/L) for microcystin spiked lake water and Super Q™ samples

				Approximate Initial MC Dosage					
	C	*DOC*	*UVA*	*No spiked MC*		*0.5 (µg/L)*		*1.0 (µg/L)*	
	(1/S)	*(mg/L)*	*(1/cm)*	*ELISA*	*PP2A*	*ELISA*	*PP2A*	*ELISA*	*PP2A*
Lake Pleasant	*1046*	*0.0475*	*0.0475*	*Blank*	*Blank*	*0.59*	*0.40*	*1.04*	*0.78*
Bartlett Lake	*665*	*0.0319*	*0.0319*	*Blank*	*Blank*	*0.43*	*0.53*	*0.94*	*0.85*
Saguaro Lake	*1974*	*0.00571*	*0.0571*	*Blank*	*Blank*	*0.47*	*0.28*	*1.52*	*0.78*
Super Q™	*NA*	*NA*	*NA*	*Blank*	*Blank*	*<0.147*	*0.28*	*0.98*	*0.64*

Note: MC concentrations were obtained by averaging the results from three wells (n = 3) for each assay. Where C is the conductance of the water sample.

Table 3.5
ELISA and PP2A results for frozen, microcystin spiked lake water and Super Q™ samples

Lake	*Sample Number*	*Unfrozen MC concentration (µg/L)*	*Frozen MC concentration (µg/L)*
Pleasant	*1*	*1.00*	*0.85*
	2	*0.96*	*0.89*
	3	*1.14*	*0.99*
Bartlett	*4*	*1.01*	*0.87*
	5	*0.94*	*0.99*
	6	*0.88*	*0.96*
Saguaro	*7*	*1.72*	*0.85*
	8	*1.31*	*0.85*
	9	*1.52*	*0.99*
Super Q™	*10*	*0.97*	*0.82*
	11	*1.1*	*0.80*
	12	*0.89*	*0.95*
Mean		*1.12*	*0.90*
Standard Deviation		*0.25*	*0.07*

Note: Frozen samples were allowed to sit in freezer at –20 degrees Celsius for two weeks. MC concentrations were obtained by averaging the results from three wells (n = 3).

Table 3.6
ELISA calibrator concentrations predicted by PP2A assay

MC Concentration (μg/L)		*Percent Difference (%)*
ELISA (C_1)	*PP2A* (C_2)	$\frac{[C_1 - C_2]}{C_1} \times 100\%$
0.16	*0.14*	*12.5*
0.16	*0.18*	*12.5*
0.16	*0.14*	*12.5*
0.5	*0.34*	*32.0*
0.5	*0.93*	*86.0*
1.6	*1.6*	*0*
1.6	*1.6*	*0*
1.6	*1.1*	*31.3*

Table 3.7
Microcystin concentration for C18 extracted samples (initial concentration 1.2 μg-MC/L)

Replicate #		MC Concentration
		(μg/L)
Sample 1		0.82
Sample 2		1.1
Sample 3		1.6
Average (μg/L)		1.2
Variance ($\mu g^2/L^2$)		0.18
Standard Deviation (μg/L)		0.42

Table 3.8
Microcystin concentrations recovery using PP2A

Sample Site	MC Concentration Average (μg/L)	
	No C18	With C18
May 2002 Bartlett	<0.05	0.00021
May 2002 Saguaro	<0.05	0.00024
May 2002 Pleasant	<0.05	0.00029
April 2002 Deer Valley	<0.05	0.00032

Table 3.9
Comparison of intracellular extraction methods

Sample Number	Microcystin concentration (μg-MC/OD_{730})		
	Direct Beadbeating	24 hrs in MeOH + Beadbeating	24 hrs in MeOH +C18 extraction
1	0.594	0.895	1.261
2	0.618	1.160	0.619
3	0.652	0.709	1.057
4	0.488	0.981	0.528
5	0.388	1.090	0.510
6	0.472	0.811	
7	0.490		
8	0.512		
9	0.580		
Mean	0.533	0.627	0.795
Standard Deviation	0.084	0.170	0.342

Note: Culture Growth = 15 days; OD_{730} = 1.05 cm^{-1}; Extracellular concentration = 0.62 μg-MC/L; Intracellular concentration = 494 ± 89 μg-MC/L

Table 3.10
Intracellular extraction of *Microcystis aeruginosa* cultures

Sample Number	Microcystin concentration (μg-MC/OD_{730})			
	24 hrs in MeOH + Beadbeating		24 hrs in MeOH	
	10 day	29 day	10 day	29 day
1	0.662	1.042	0.437	>1.036
2	0.732	1.076	0.380	>1.036
3	0.634	0.991	0.408	>1.036
4	0.915	0.762	0.775	0.949
5	0.887	0.758	0.648	0.990
6	0.859	0.275	0.592	0.966
7		0.242	1.056	0.680
8		0.275	0.944	0.874
9			0.958	0.829
Mean	0.782	0.926	0.689	0.881
Standard Deviation	0.121	0.154	0.257	0.116

Note: Culture Growth = 10 days; OD_{730} = 0.142 (1/cm); Extracellular conc. = > 2.5μg-MC/L; Intracellular conc. = 56 μg-MC/L. Culture Growth = 29 days; OD_{730} = 1.81 (1/cm); Extracellular conc. = >25 μg-MC/L; Intracellular conc. = 838 μg-MC/L

Table 3.11
Optical density and related biomass for *Microcystis aeruginosa* culture

Culture Growth	OD_{730}	Biomass Concentration	
(days)	(1/cm)	(mg-biomass/L)	μg-biomass/OD_{730}
21	0.95	0.356	0.75
21	0.98	0.282	0.58
41	2.22	0.940	0.85
41	2.50	1.270	1.02

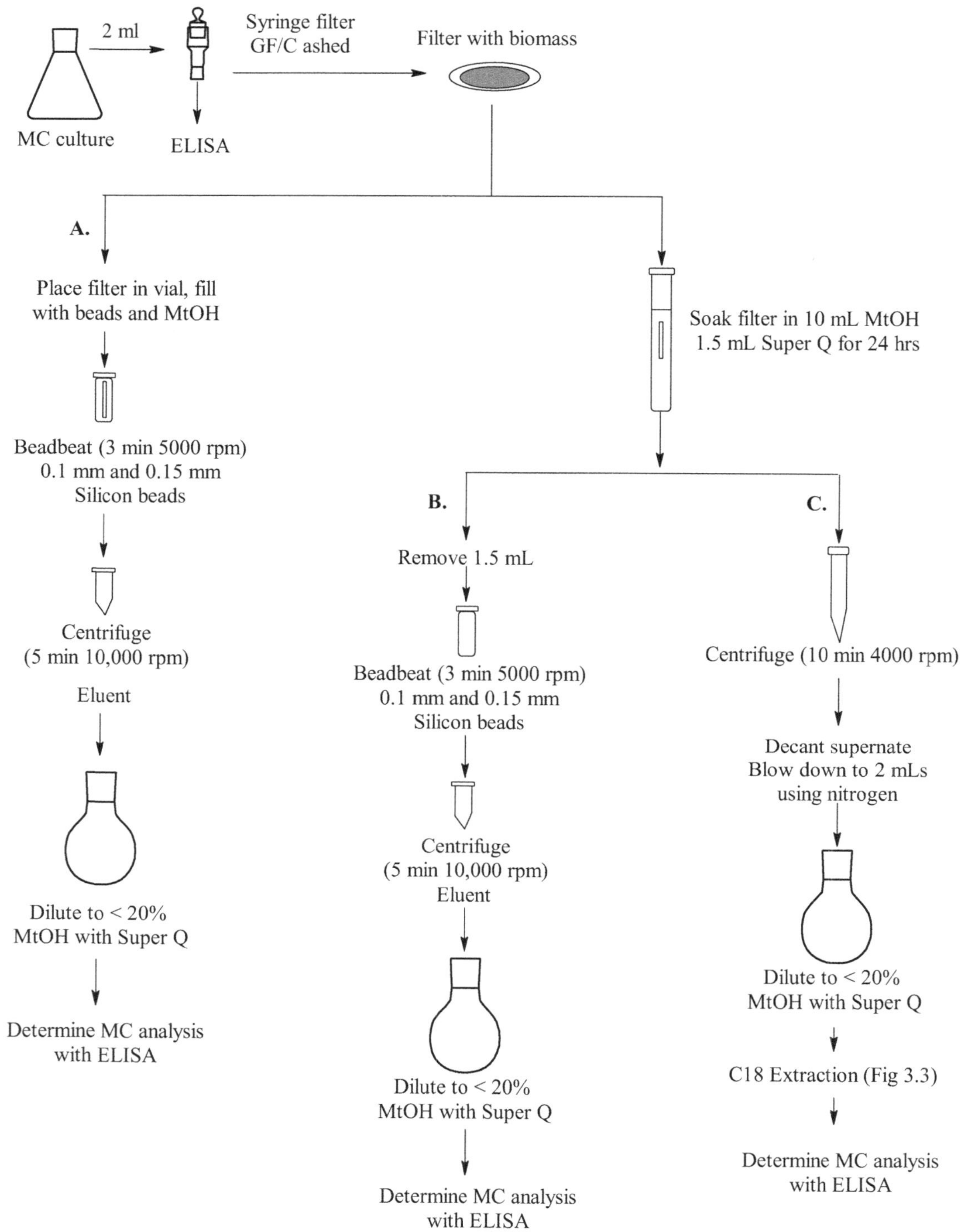

Figure 3.1 Schematic of intracellular extraction methods A, B, and C.

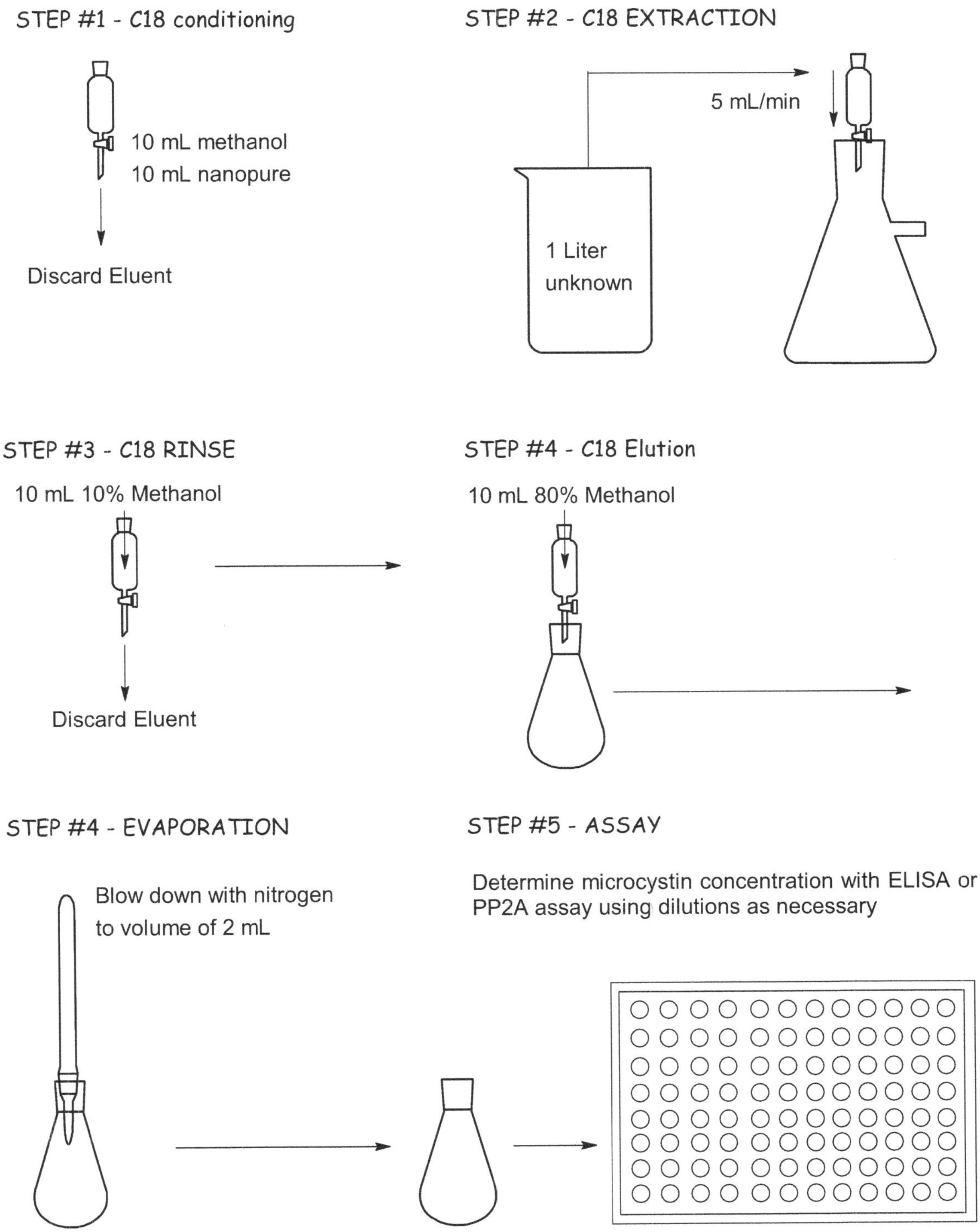

Figure 3.2 Schematic of solid phase extraction procedures

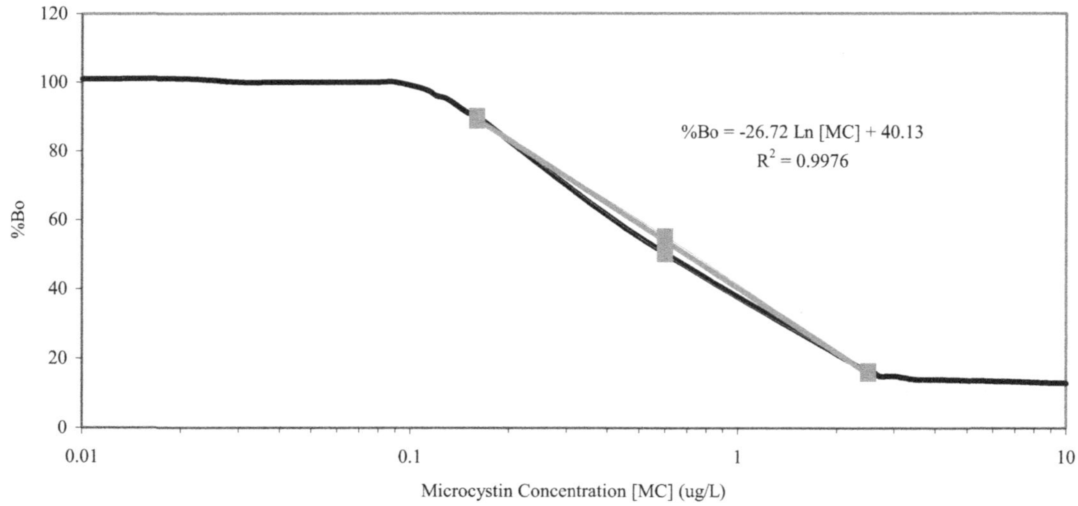

Figure 3.3 Relationship between optical density (% Bo) and microcystin concentration for ELISA standard curve development

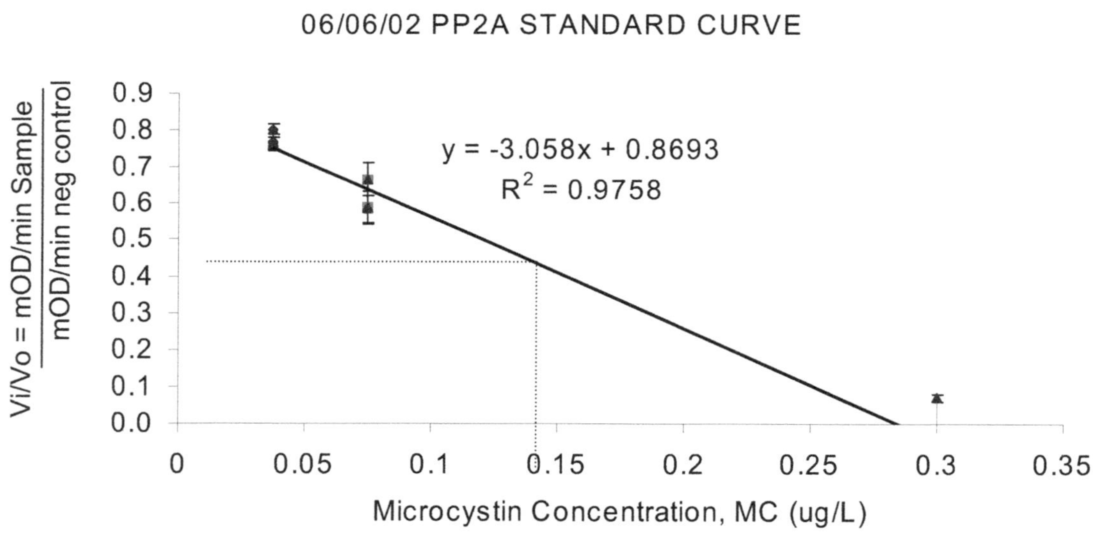

Figure 3.4 PP2A calibration curve using enzyme #1

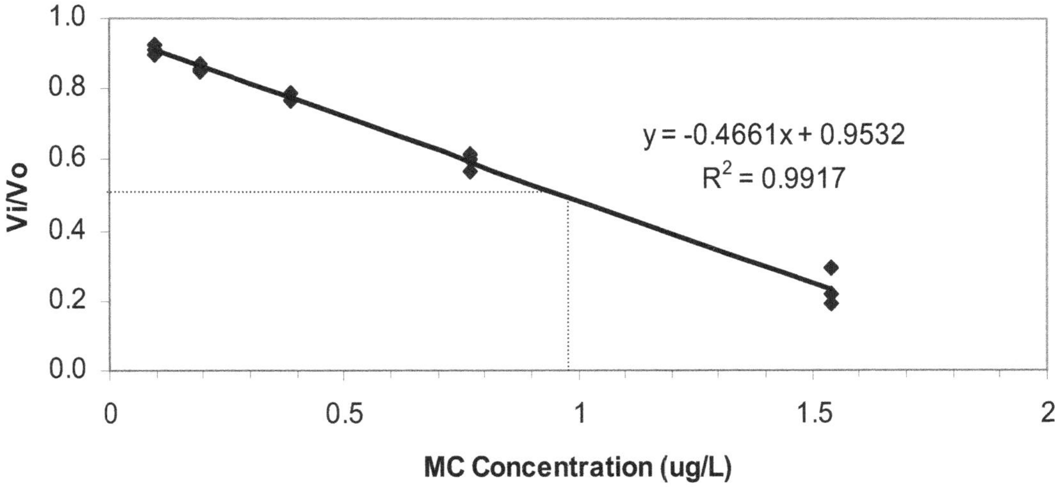

Figure 3.5 PP2A calibration curve using enzyme #4

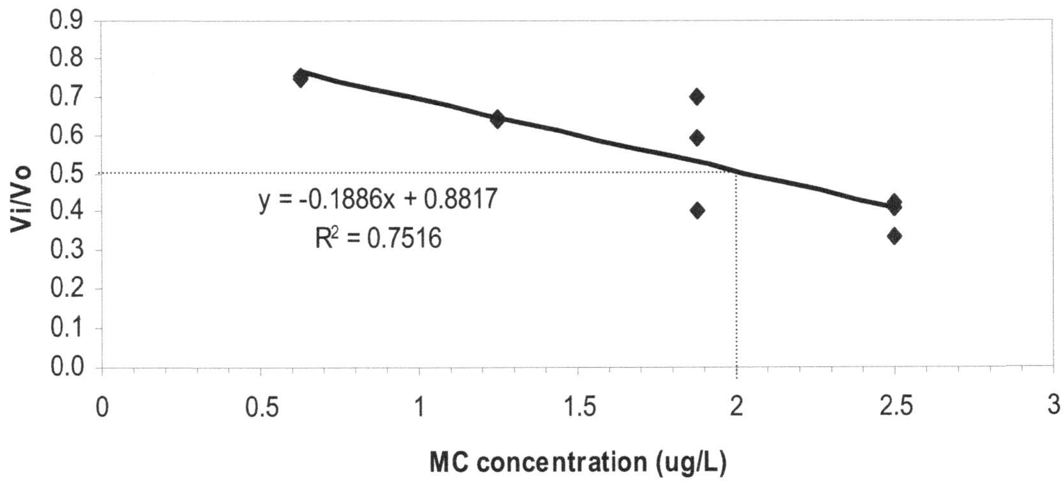

Figure 3.6 PP2A calibration curve using enzyme #5

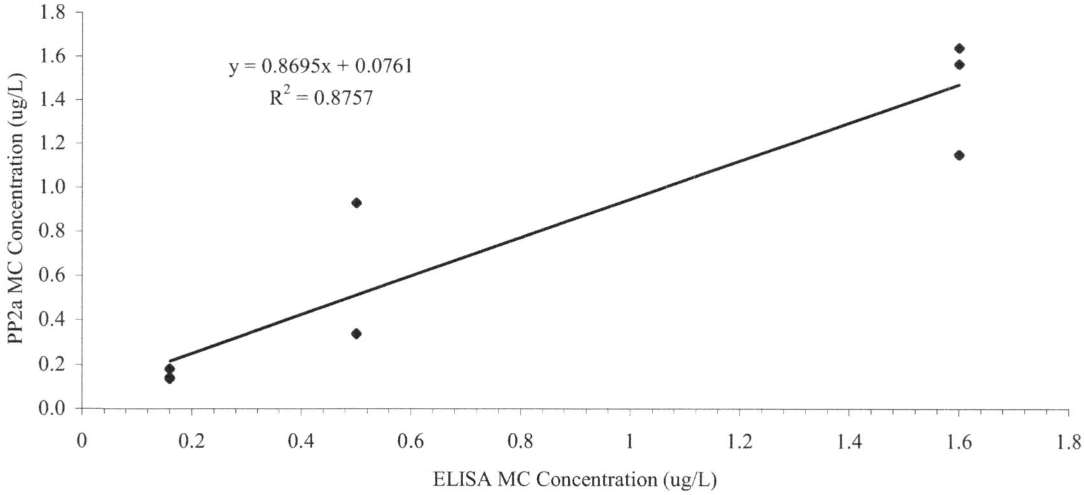

Figure 3.7 ELISA calibrations versus PP2A measured concentrations

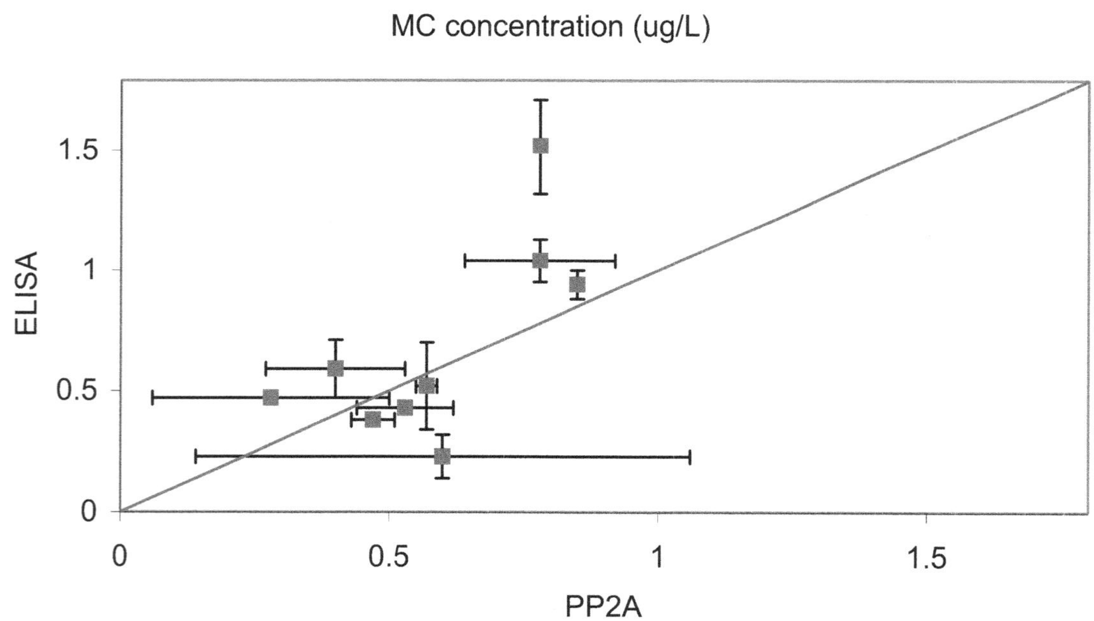

Figure 3.8 Comparison ELISA and PP2A microcystin concentrations

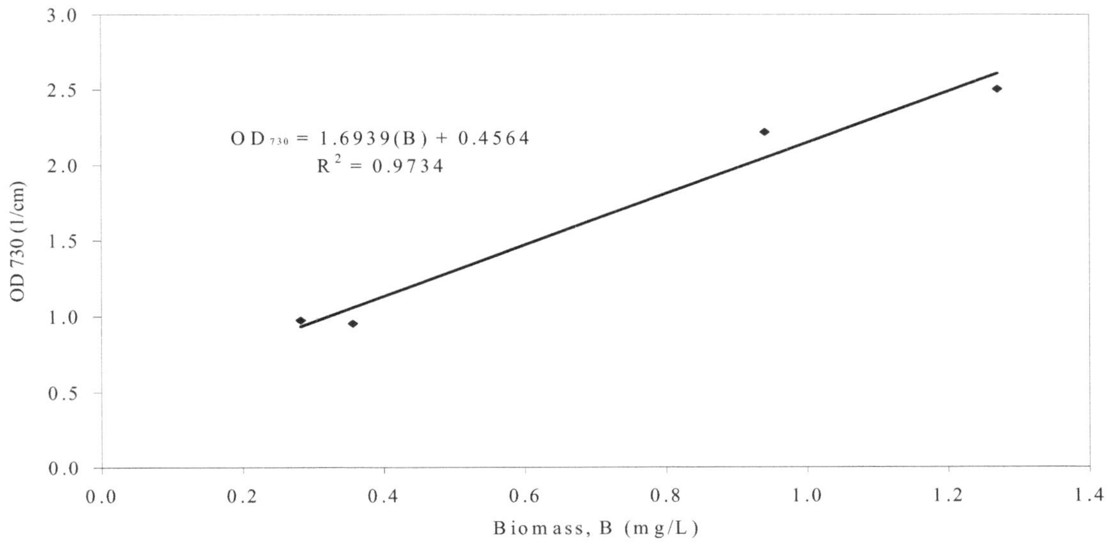

Figure 3.9 Relationship between optical density (OD_{730}) and *Microcystis aeruginosa* biomass

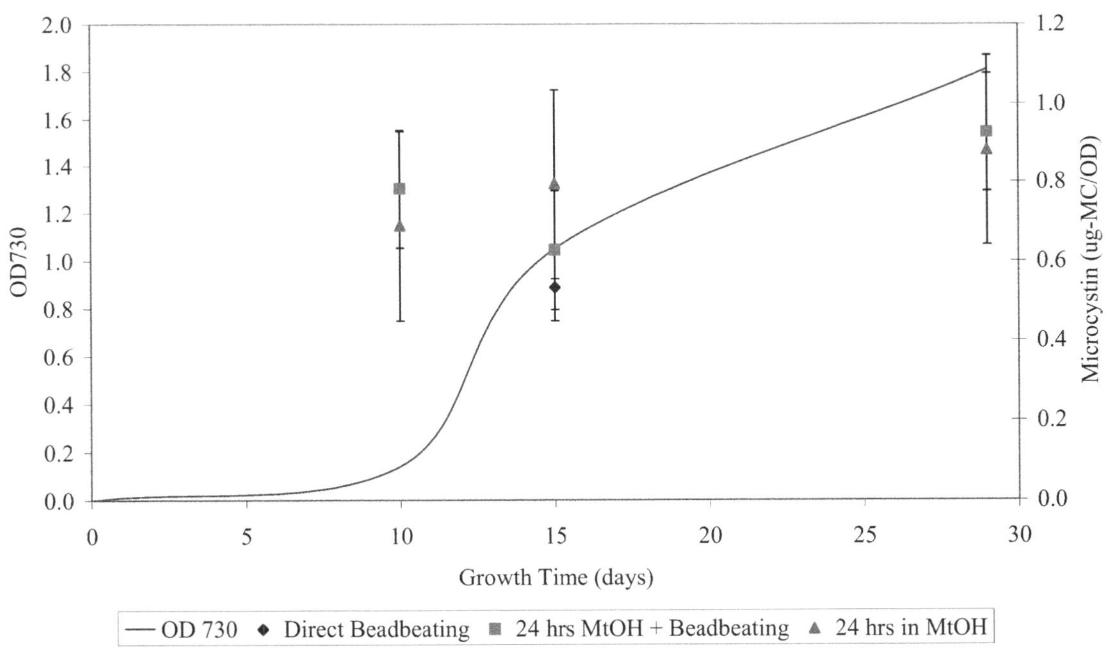

Figure 3.10 Optical density and intracellular microcystin change over time in *Microcystis aeruginosa* culture

CHAPTER 4
ALGAE METABOLITES IN RAW AND TREATED DRINKING WATERS

This chapter discusses the occurrence of algae that produce T&O compounds in the valley. Production of eight T&O compounds are explored. Occurrence and treatment of T&O and toxin algae metabolites are evaluated.

BACKGROUND

Cyanobacteria, or blue-green algae, are found in a variety of environments. Habitats include hot springs, terrestrial, freshwater, and marine environments. Water quality problems created by cyanobacteria include bloom development, T&O problems, and toxin production. Cyanobacteria are present in the environment as plankton (free floating), periphyton (attached to surfaces), or mats. Historically, there has been some debate over the classification of cyanobacteria as algae or as bacteria. Yoo et al. 1995 explained that cyanobacteria are prokaryotic, and have no organized structures including plasmids and mitochondria; thus, more closely resembling bacteria than eukaryotic algae. Cyanobacteria make up one of three groups of pigmented bacteria. The other two groups are purple and green sulfur bacteria. However, unlike purple and green sulfur bacteria, cyanobacteria produce oxygen during photosynthesis. Phycologists view cyanobacteria as prokaryotic algae, because their pigments and photosynthetic mechanisms more closely resemble those of true algae than those of photosynthetic green and purple sulfur bacteria. Often cyanobacteria are classified as a separate, distinct group. Skulberg et al. 1993 further discuss the classification debate.

Cyanobacteria range from unicellular forms to colonial unicellular and filamentous forms. The size of the cells varies with individual organisms. They range from only a few micrometers in diameter, to much larger sizes (Yoo et al. 1995). For example, the individual cells in the filaments of *Oscillatoria princeps* may be as large as 60 mm in diameter. This size represents one of the largest cells among the prokaryotes (Yoo et al. 1995).

Cyanobacteria contain pigments (phycobilins) that reflect both blue and green wavelengths, and are often referred to as blue-green algae. One class of the phycobilins is phycocyanin, which is responsible for the blue appearance of the algae. Phycocyanin traps light energy in the red band of the visible spectrum, and then transfers this energy to the chlorophyll-a pigment that is used in photosynthesis. Chlorophyll-a absorbs a blue wavelength for energy and reflects green. Cyanobacteria use the blue and green pigments to trap light at both extremes of the visible spectrum. Thus blue-green algae utilize deep-penetrating red wavelengths in addition to the blue wavelengths near the surface, allowing cyanobacteria to continue photosynthesis relatively independent of their position in the water column (Yoo et al. 1995). However, not all cyanobacteria are blue-green in color. Cyanobacteria colors range from olive-green, gray-green, and yellow-brown to purplish and red.

Gas vacuoles allow some species of cyanobacteria to regulate their depth in the water column. Gas vacuoles are comprised of vesicles that fill and collapse in a cyclic process regulated by the accumulation and synthesis of photosynthetic by-products. Advantages of depth regulation include improved photosynthesis through movement to light, and access to nutrients throughout the water column.

Maier et al. 2000 discussed the significance of cyanobacteria in the aquatic environment. Cyanobacteria are primary producers in the microbial planktonic community. Organic carbon and energy produced from cyanobacteria photosynthetic processes are transferred to other organisms as part of the food web. Cyanobacteria are also consumed by zooplankton, which are then consumed by larger organisms as a part of the grazing food chain. Additionally, more than 50% of the carbon fixed during photosynthesis of the phytoplankton is released into the surrounding water as dissolved organic matter (DOM). Bacterioplankton utilize this DOM as part of the microbial loop. The bacterioplankton then either mineralize the organic carbon into carbon dioxide, or produce new biomass. Thus cyanobacteria play a vital role in aquatic environments in primary production (Maier et al. 2000).

Cyanobacteria are also an essential component in microbial mats. Maier et al. 2000 explain that cyanobacteria are the most important source of organic carbon in a microbial mat. Cyanobacteria access sunlight by occupying the top layer of mats. The photosynthetic activity of the blue-green algae creates an oxygen rich environment for additional microbial growth. Cyanobacteria work with sulfur-oxidizing bacteria to complete a sulfur and nitrogen cycle. Stal, 1995 explained that cyanobacteria, sediment, and organic debris form the surface layer of the mat preventing excessive ultraviolet radiation from penetrating the mat. Microbial mats are often the only life forms present in very extreme environments.

Cyanobacteria impact the environment in several ways. Blue-green algae blooms lead to undesirable tastes and odors, as well as potential toxin release. Cyanobacteria blooms are defined as prolific planktonic algae in aquatic habitats housing millions of cells per liter (Carmichael, 1992). Potential toxicity of a cyanobacteria bloom has been connected to bloom size. Yoo et al. 1995 noted that the likelihood of serious cyanobacteria toxicity occurring increases with increasing cyanobacteria mass. Nutrient concentrations, water temperatures between 15-30° C, and pH values between 6 and 9, provide conditions favorable to bloom development. The highest concentration of cyanobacteria typically occurs at a depth between 2-9 meters (Maier et al. 2000). The fluctuation of T&O compounds 2-methylisoborneol (MIB) and geosmin in the City of Phoenix water supply may also provide an indicator of MC blooms. However, three of the genera of cyanobacteria that are dominant T&O producers in the COP water supply may also release MC. These genera include *Oscillatoria*, *Anabaena*, and *Microcystis*. Research in Alberta, Canada found no relationship between MIB and geosmin and MC concentrations (Hrudey et al. 1993). Aquatic environments with warm temperatures, and moderate pH values should be monitored for toxins released by cyanobacteria blooms.

Algal blooms are often the source of undesirable T&O compounds in drinking water. T&Os common in drinking waters include earthy/musty, sour, woody/hay, metallic, chlorinous, fishy, medicinal and chemical (Suffet et al. 1993; Suffet et al. 1995). Geosmin and MIB are two common algal metabolites producing T&O compounds identified in the COP water supply (Bruce et al. 2002). Geosmin has an earthy odor and MIB a musty odor. Detectable odors for MIB and geosmin are 29 ng/L for MIB and 10 ng/L for geosmin (Wnorowski, 1992). Water treatment plants that utilize surface water commonly have T&O problems caused by algal blooms. In a national survey of public water suppliers; 46% reported algal growth and 22% reported earthy-moldy T&O in their source waters (Suffet et al. 1993). β-cyclocitral (threshold of 5 ng/L) is another algal-produced T&O compound described as fresh grass, hay/woody, and tobacco-like odors (Young et al. 1999). β-cyclocitral and geosmin have been identified in the same water body. A significant T&O episode in Appleton, Wisconsin identified geosmin and β-cyclocitral as the culprit compounds (Young et al. 1999). Algal T&O problems are typically caused by phytoplankton, but Metropolitan Water District of Southern California research indicated that periphyton contributes to T&O problems as well (Izaguirre et al, 1995).

DESCRIPTION OF ALGAE PRESENT IN WATER SUPPLY

Production of toxins harmful to humans and animals have been associated with cyanobacteria blooms. Cyanotoxins have been linked to liver damage, gastrointestinal disturbances, and neural damage. The cyanobacterial hepatotoxin MC-LR was attributed to the lethal acute poisonings of humans in Caruaru, Brazil in 1996 when MC contaminated water was used in dialysis treatment (Pouria et al. 1998). A study in China found a correlation between the occurrence of primary liver cancer and drinking water heavily infested by toxic cyanobacteria (Ueno et al. 1996). Water supplies prone to algal and cyanobacteria blooms should be monitored to detect cyanotoxins. Based on the toxicity of MC, the WHO has set a limit of 1 μg/L for MC in drinking water (Chorus et al. 1999). The 1994 Canadian guideline for MC-LR (most common form of MC found in water supplies) specifies a maximum acceptable concentration of 0.5 μg/L of total MC-LR or 1 μg/L of total MC in drinking water (Carmichael, 2000). Carmichael, 2000 tested 677 water samples from water utilities in both the United States and Canada for MC from June 1996 through January 1998. Of these samples 80% were positive for MC, with 4.3% of the samples at MC concentrations greater than 1 μg/L. Based on the acute and chronic risks of MC, MC and other cyanobacteria toxins are listed on the contaminant candidate list by the US Environmental Protection Agency (EPA, 2002).

Currently it is unclear why cyanobacteria release toxins. Cyanobacteria respond differently to changing environmental conditions (Rapala et al. 1993). Jones et al. 1993 concluded that light and nitrogen availability had the greatest affect on toxin production. They further noted that toxin production varies over time and may be explained by a combination of environmental factors, including ecological changes of dominating cyanobacteria over time.

Microcystis is the number one toxin producing genera of cyanobacteria (Maier et al. 2000), but is not the only genus that produces the toxin MC. Table 4.1 summarizes confirmed MC producers. The most often listed producers include *Microcystis aeruginosa, Aphanizomenon flos-aquae, Anabaena flos-aquae, Oscillatoria agardhii,* and *Lyngbya* spp. (Yoo et al. 1995). The toxin MC may be present in several different forms. These variants include MC-LR, -LA, -YA, YM, -RR, and –YR.

Several cyanobacteria were identified and isolated as part of the COP/ASU T&O project conducted over the past years (Table 4.2). Potential MC producers identified include *Anabaena* and *Oscillatoria*. A large variety of cyanobacteria exist in reservoir phytoplankton. Table 4.2 is not inclusive of all observed cyanobacteria. However, the identified and isolated species allow characterization of the planktonic make up of the COP source waters, and indicate the presence of potential MC producers.

OCCURRENCE OF MIB AND GEOSMIN IN WATER SUPPLY

Lakes

Figure 4.1 shows the long-term trends in MIB for the three reservoirs serving metro-Phoenix with raw drinking water. Saguaro Lake exhibits MIB production every year, while Bartlett and Pleasant Lakes appear to experience higher MIB events only two out of the four years of record. MIB levels are higher in the upper layers of the reservoir (epilimnion). Upon thermal destratification (September for most reservoirs in 2003), MIB concentrations were uniform with depth. Water withdrawn from the reservoirs during and immediately after thermal destratification will cause a change in MIB and other water quality parameter being released downstream. MIB concentrations quickly decrease in the months after destratification due partially, over the course of 1-2 weeks, to dilution, but mainly through biodegradation of the MIB by bacteria naturally occurring in the reservoirs. Geosmin concentrations generally paralleled MIB concentrations, but geosmin concentrations were lower than MIB concentrations in the reservoirs.

Canals

MIB production in the canal system occurred noticeably only in the Arizona Canal during the warmer months of 2003 (Figures 4.2 and 4.3). Two regions of MIB and geosmin production were observed. These included the reach from Highway 87 to the siphon near Hayden Road and between 24th Street and 29th Avenue. Over the same time period SRP funded a genetic fingerprinting project to identify culprit periphytic algae that grow on the canal walls (Hu, Sommerfeld, Westerhoff). Culprit algae are present and genetic analysis can be an early warning approach for timing of chemical or mechanical treatments.

In both sections of the Arizona Canal blue-green algae were the suspected cause of MIB and geosmin production along the length of the canal. However, in late 2003 sediment samples collected from the Arizona Canal between Pima Road and Hayden Road indicated the presence of actinomycetes, some of which can also produce MIB and geosmin. Above the siphon on the Arizona Canal (near Hayden Road) rich organic sediment was present in the canal and would be ideal for actinomycete growth. Above Pima Road the canal sediment was limited to more inorganic materials (sand, gravel), but actinomycetes were still present.

The geometry of these two canal sections are very different, and hence the ability to treat the canals also differs. The lower reaches of the Arizona Canal have the walls cleaned mechanically using the SRP brushing device. The upper reaches of the Arizona Canal has steeper side banks and deeper water and is more amenable to chemical algaecide treatments than brushing.

Water Treatment Plants

Data for the influent and effluent T&O levels at the different water treatment plants (WTPs) is presented in Figure 4.4 for each WTP. A negative removal indicates production within the WTPs (e.g., South Tempe WTP and Union Hills). The percentage removal is a function of influent MIB concentration, so the percentage removals are only intended to provide a relative assessment of each plants ability to remove (or produce) MIB.

Tables 4.3 and 4.4 summarize the WTP raw and finished (Fin.) water MIB concentrations based upon different MIB concentration ranges. The number of samples is based on approximate monthly sampling January through June and weekly sampling July through December; finished water samples were not collected when facilities were off-line. Union Hills WTP and South Tempe WTP had very little MIB in raw (or finished water). Greenway WTP had significant levels of MIB in raw water, but the ozone-GAC biofiltration process removed the MIB very well. Phoenix's 24th Street, Deer Valley, and Val Vista WTPs did not have to control

MIB when the raw water was < 5 ng/L. PAC addition at these WTPs kept MIB concentrations below 30 ng/L, but PAC doses were not high enough to consistently keep finished water levels below 10 ng/L. The North Tempe WTP performed similar to Phoenix WTPs, but tended to add a little more PAC and had more days with lower MIB concentrations.

PRODUCTION OF NON-MIB AND NON-GEOSMIN EARTHY-MUSTY ODORS BY ALGAE

Non-MIB/Geosmin T&O compounds in isolated algal cultures were identified by SPME coupled with GC-MS tests conducted at COP laboratories. The cultures examined were from the COP reservoirs, rivers and canals. Isolates were cultured in BG-11 growth media to obtain a highly concentrated solution of potential T&O compounds. The cultures were allowed to grow until the culture was producing an odor detectable by the human nose. The BG-11 algal mixture was then filtered using a GF/C syringe filter to remove cellular material that interferes with GC-MS analysis. The eluent was next delivered to the COP laboratory for SPME analysis to determine the T&O culprit compound. The compounds examined in this testing process were obtained from the "Taste and Odor Wheel" [18] and include: cis-3-hexen-1-ol, cis-3-hexenyl Acetate, 2-Isobutyl-3-Methoxypyrazine (IBMP), 2-Isopropyl-3-Methoxypyrzaine (IPMP), Methylisoborneol, β-cyclocitral, Geosmin, and 2,4,6-Tribromoanisole.

The results of the non-MIB or Geosmin T&O compounds are presented in Table 4.5. As indicated in the table, all but one of the cultures (*Leptolyngya*) were positive for β-cyclocitral. Additionally, the majority of the cultures did not contain any other compounds at concentrations greater than the detection limits of SPME. However the *Plankothrix* was positive for 2,4,6-tribromoanisole, and the *Lyngbya* was positive for c is-3-hexen-1-ol. It is important to note that these results should not be interpreted quantitatively. The algal cultures were allowed to grow for various amounts of time, which resulted in varying compound concentrations. The purpose of this test was to determine the presence of T&O culprit compounds; therefore, results should be interpreted as positive or negative for the presence of a particular compound. Based on these results, the presence of β-cyclocitral in the COP water supply may play a significant role in T&O episodes occurring in the Phoenix metro area. Additionally, isolated cyanobacteria were also tested for MC. Microcystin was not found in any of the samples based on analysis using PP2A.

β-cyclocitral can be produced by algae in the COP water supply and may play a significant role in the T&O occurrence. Of the twenty-one isolated cultures examined using SPME, twenty samples were positive for β-cyclocitral. Future work concerning β-cyclocitral in the COP water supply should include removal of β-cyclocitral by the water treatment processes, specifically PAC, and monitoring of production trends to determine critical T&O production periods throughout the year.

OCCURRENCE OF MICROCYSTINS IN WATER SUPPLY

Microcystin concentrations in the Phoenix water supply were monitored using both ELISA and PP2A assays. Concentrations were undetectable from September 2001 through May 2002. In June 2002 the MC concentration in the reservoirs increased and detectable concentrations (> 0.147 μg-MC/L) persisted through the end of October. Canal and WTP influent concentrations were monitored during this time frame. Several factors contribute to the production of MC. This chapter presents the observed change in MC concentration and potential correlation between MC concentration and algal biomass (chlorophyll-a), T&O compounds (MIB and geosmin), nutrients (phosphorous and nitrogen), organic carbon and light penetration (secchi disc depth) in the reservoirs over time.

TEMPORAL VARIATION IN MICROCYSTIN CONCENTRATION

The MC concentration in the Phoenix water supply varies with time. Figure 4.5 illustrates the MC fluctuations over the past 14 months, and Figure 4.6 shows the MC fluctuations in the canals and the 24th Street WTP for August and September 2002. Values less than the assay detection limit (0.147 for ELISA, 0.05 or 0.24 μg-MC/L for PP2A) are represented as zero. Deer Valley WTP is located at the end of the Arizona Canal and the MC concentration of the influent was monitored through part of the year to indicate periphytic MC production in the canal (Figure 4.7). The Deer Valley MC concentration closely follows the lake concentrations, indicating little or no periphytic MC contribution. The MC concentration was also measured in the Deer Valley WTP immediately following PAC addition during August 2002. In general the MC concentration decreased after contact time with PAC. Hence PAC used for T&O compound absorption did reduce MC concentration. On August 28, 2002, the MC concentration was tracked through the Water Treatment Plant process (Table 4.6). The influent concentration was 0.32 μg-MC/L, the concentration following PAC was < 0.1 μg-MC/L for this PP2A assay, and the effluent concentration was 0.11 μg-MC/L. The effluent concentration may have increased through release of intracellular MC following chlorine addition. The concentration did not appear to vary much following PAC and chlorine addition. At low concentrations of MC, this data is inconclusive. Future research should include the monitoring of MC through WTPs in the metro area.

Microcystin concentrations in the reservoirs varied with time. The highest dissolved MC levels were measured in July. Samples were frozen for one to four weeks prior to conducting assays from September 2001 through May 2002. MC concentrations may have been higher prior to freezing of the samples. It is therefore possible that MC concentrations were greater than assay detection limits prior to June 2002. Using C18 extraction techniques, MC at the part per trillion level was measured in September of 2001 and again in May of 2002. Hence MC is potentially in the water supply year round, but did not reach detectable levels via ELISA and PP2A assays until June. Dissolved MC levels were below the WHO action level of 1 μg-MC/L for total MC throughout the monitoring period. However, based on the dynamic nature of algal blooms, MC concentrations should continue to be monitored in the future.

Canal samples were monitored for MC throughout August and September of 2002 (Figure 4.6). The purpose of this monitoring was to determine if periphytic algal growth contributed to MC in the water supply. Microcystin concentrations measured in R12 samples include a combination of Saguaro and Bartlett Lake water. The R12 MC concentrations closely follow the lake concentrations, indicating no periphytic MC contribution. Lake Pleasant water flows through site R11 the CAP canal. Lake Pleasant had concentrations of 0.12 μg-MC/L in both August and September. R11 concentrations fluctuated around this range. A MC sample was also taken below the cross connect where CAP and SRP canal water are combined (R13). Based on the flow rates in these two canals on the day of sampling, the R13 sample showed no significant increase or loss of MC. The canal samples indicate no MC contribution by periphytic algae. Microcystin concentrations were also monitored in the 24th Street WTP. As indicated in Figure 4.6, the MC concentration in the 24th Street WTP decreases slightly following PAC addition. The MC concentration in the Union Hills WTP (located on the CAP downstream of Lake Pleasant) was also measured on August 8th and 28th, 2002. Measured concentrations were 0.24 and 0.12 μg-MC/L, respectively. These concentrations are similar to those measured in Lake Pleasant during the same time period (0.58 to 0.12 μg-MC/L), again indicating little or no periphytic MC contributions.

MICROCYSTIN PREDICTORS

Several factors may contribute to the production of MC. The following water quality parameters have been monitored for use in predicting MIB/geosmin production:

- Algal biomass (chlorophyll-a)
- T&O compounds (MIB/geosmin)
- Nutrients (total and dissolved phosphorous and nitrogen)
- Light penetration (secchi disc depth)
- Particulate organic carbon (POC)

Correlations between MC and chlorophyll-a, MIB, phosphorous (total and dissolved), nitrogen (total and dissolved), secchi disc depth, and POC were determined by trendlines developed using Excel. The data was linearly fit and the corresponding R^2 value was used to determine the extent of data correlation. An R^2 value less than 0.5 was selected to indicate insignificant correlation between variables. Using this criterion no linear correlation was found between MC concentration and chlorophyll-a, MIB, nutrients, light penetration and POC.

The change in MC concentration overtime peaked slightly earlier than the maximum MIB concentration, but loosely followed a similar temporal pattern (Figures 4.8 - 4.10). This relationship may be explained ecologically. Microcystin producing cyanobacteria may flourish in the water column during June under specific pH, temperature, nutrient and light conditions. Then as the matrices change, the MIB-producing cyanobacteria begin to dominate the column.

Microcystin concentrations peaked one to two months following peak light penetration in the reservoirs measured by secchi disc depth (Figures 4.11 - 4.13). Periods of large light penetration decreases often indicate algal blooms in reservoirs. Microcystin concentrations appear to increase as light penetration drops following peak penetration. Cyanobacteria may be flourishing during these periods of light penetration drops, and then as the cells die, MC is released and MC concentrations increase.

Changes in MC concentration over time were compared with chlorophyll-a concentrations in Bartlett and Saguaro Lakes (Table 4.7). Chlorophyll-a data was available for July through October 2002. Lake Pleasant showed a large increase in chlorophyll-a following the peak MC concentration. The increasing chlorophyll-a values may indicate an increase in competing cyanobacteria and thus a decrease in MC producing algae. Bartlett and Saguaro Lake MC and chlorophyll-a temporal data showed no distinct trends. The MC concentration decreased at one point as chlorophyll-a decreased and then continued to decrease when chlorophyll-a concentrations increased. Microcystin producers are only a small part of the total algae composition of the reservoirs and therefore MC production is not readily explained by chlorophyll-a data.

Total MC concentrations were determined for the reservoirs during August and October of 2002. Table 4.8 provides a summary of dissolved and total MC concentrations in Lake

Pleasant, Bartlett and Saguaro Lakes. The total MC concentration is just larger than the dissolved concentration. Very little MC in the samples was contained in the cells at the time of sampling. In October the MC concentrations within the cells at the time of sampling, were one order of magnitude greater than those measured in August. Based on similar total and dissolved MC concentrations, samples were probably not taken from a MC producing bloom. In future monitoring dissolved MC concentrations are expected to be similar to total concentrations unless samples come directly from a bloom.

The amount of MC per Optical Density at 730 nm (OD_{730}) was calculated for the October samples. These concentrations were > 1.52 μg-MC/OD_{730} in Lake Pleasant, 0.076 μg-MC/OD_{730} in Bartlett Lake, and 0.460 μg-MC/OD_{730} in Saguaro. The MC mass per OD_{730} values indicate that a larger percentage of the biomass in the Lake Pleasant sample includes intracellular MC than in Saguaro and Bartlett at the end of October 2002. Based on intracellular and extracellular concentrations, Lake Pleasant appears to be more prone to MC production than Bartlett and Saguaro Lakes.

POWDER ACTIVATED CARBON ABSORPTION

PAC used for T&O compound removal absorbs MC. The percent remaining of MC toxin after absorption by PAC was determined based on PP2A assay results. The percent remaining was defined as C/Co multiplied by 100%. Where C is the final concentration (μg-MC/L) after contact with PAC, and Co is the initial concentration (μg-MC/L). The percent remaining MC was then plotted versus PAC dosage to compare MC absorption by three PAC brands.

AquaNuchar, Norit HDB and Norit 20B absorbed MC (Figure 4.14) The AquaNuchar removed forty percent of the MC at a dosage of 1 mg-PAC/L, the Norit HDB removed approximately fifty percent, and the Norit 20B removed ten percent. At a PAC dose of 5 mg/L, the AquaNuchar had removed approximately ninety percent of the MC, while only a slight increase in percent removed was observed for the Norit HDB and the Norit 20B. As the PAC dosage increased, the performance of the Norit HDB and Norit 20B became more similar. By a dose of 8 mg-PAC/L, the AquaNuchar had removed sufficient MC to have a concentration less than the WHO limit of 1 μg/L. A dosage of 20 mg-PAC/L of the Norit HDB achieved a similar result. A dose of 50 mg-PAC/L of the Norit 20B was required to obtain a concentration less than 1 μg-MC/L. At the lowest PAC dosage, the AquaNuchar provided the greatest MC absorption.

The material used to make the PAC provides different absorption capabilities. Both the Norit HDB and the Norit 20B brands of PAC are made with coal. The AquaNuchar is a wood-based PAC. The wood-based PACs have been shown to be better absorbents of MC than the coal based PACs (Donati et al. 1994; Mohammed et al. 1999). Donati et al. (1994) further explained that this difference in absorption capability was attributed to mesopores volume. As anticipated from the literature, the wood based AquaNuchar showed the greatest absorption capacity for the MC. The AquaNuchar removed sufficient MC to achieve acceptable concentrations (< 1 μg/L) at a dose of 8 mg-PAC/L. This carbon dosage (8 mg-PAC/L) is less than the PAC dosage applied in the metro area during T&O episodes (5 - 50 mg-PAC/L) and therefore should absorb MC present in the water as well as T&O compounds. The Norit HDB and Norit 20B also removed sufficient MC (< 1 μg-MC/L), but at PAC dosages greater than those typically employed at the Phoenix metro water treatment plants. Use of the AquaNuchar or other suitable wood-based PACs is recommended for future MC and T&O compound removal.

CHLORINE OXIDATION

Low levels of chlorine are capable of oxidizing MC with sufficient contact time. Results of the residual chlorine test are presented in Figure 4.15. Microcystin concentrations after a 14-day contact time were determined using PP2A and ELISA assays. As seen in Figure 4.15, application of a low chlorine dose (0.1 mg-chlorine/L) oxidized 96% of the initial 13 μg-MC/L. For dosages less than 0.1 mg-chlorine/L, no oxidation of MC appeared to occur. Glassware and water both have some chlorine demand and may have used up the available chlorine in the 0.02 and 0.05 mg-chlorine/L samples. The COP currently uses free chlorine to provide residual disinfection in the water supply. Based on these results, typical residual chlorine dosages from a WTP should be sufficient to oxidize MC in the drinking water.

SUMMARY

Monitoring of MC concentrations over time indicated:

- Saguaro Lake exhibits MIB production every year, while Bartlett and Pleasant appear to experience higher MIB events only two out of the four years.
- MIB levels are higher in the upper layers of the reservoir (epilimnion).
- MIB production in the canal system occurred noticeably only in the Arizona Canal during the warmer months of 2003.
- β-cyclocitral can be produced by algae in the COP water supply and may play a significant role in the T&O occurrence.
- July was the peak month for MC production with the highest dissolved concentration in Lake Pleasant of 0.53 μg-MC/L.
- Microcystin was measured in the reservoirs from June through October 2002 at levels greater than 0.147 μg-MC/L, the detection limit for ELISA.
- Continued monitoring of MC in the reservoirs is recommended based on the dynamic nature of cyanobacteria blooms.
- No periphytic MC contribution was observed in the canals.
- The MC concentration measured in the Deer Valley WTP immediately following PAC addition decreased. Hence PAC used for T&O compound absorption did reduce MC concentration. Additional research should be done to determine effectiveness of WTP processes on MC removal.
- No linear correlations ($R^2 > 0.5$) were found between MC concentration and chlorophyll-a, MIB, phosphorous, secchi disc depth, and POC.
- Temporally:
  - MC peak production precedes MIB peak concentrations by one month.
  - MC follows peak light penetration (maximum secchi disc depth) by one month.
  - MC concentrations do not appear to follow chlorophyll-a temporal trends.
- AquaNuchar, a wood based PAC, absorbs MC at dosages currently used by the COP for T&O removal (5 - 50 mg-PAC/L).
- Low levels of chlorine (0.1 mg-chlorine/L) oxidize MC given sufficient contact time.

Table 4.1
Summary of identified microcystin producers

Organisms Producing Only Hepatotoxic Microcystins or Nodularians	*Organisms Producing Both Hepatotoxins and Neurotoxic Anatoxins or PSPs*
Microcystis aeruginosa Kütz.	Anabaena circinalis *Rabenh.*
Microcystis viridis *(A. Br.) Lemm*	Anabaena flos-aquae (Lyngb.) *Brėb.*
Microcystis wesenbergii *Kom.*	Anabaena spiroides *var.* contracta *Kleb.*
Nodularia spumigena *Mertens*	Anabaena variabilis *Kütz.*
Nostoc rivulare *Kütz.*	Aphanizomenon flos-aquae *(L.) Ralfs*
Oscillatoria acutissima *Kuff*	Oscillatoria agardhii *Gom..*
Oscillatoria agardhii/rubescens *group*	
Oscillatoria nigro-viridis *Thwaites*	

Source: Summarized from Yoo et al. 1995

Table 4.2
Comprehensive list of taxa (excluding diatoms) observed in phytoplankton in the City of Phoenix water supply system

Chlorophyta (30)	*Chlorophyta (cont.)*	*Cyanophyta (17)*	*Other (10)*
Ankistrodesmus sp.	Pandorina sp.	Anabaena sp.	Ceratium sp.
Chlamydomonas sp.	Pediastrum sp.	Aphanothece sp.	Cryptomonas sp.
Chlorella sp.	Platymonas sp.	Chroothece sp.	Dinobryon sp.
Chlorococcum sp	Pyramimonas sp.	Cylindrospermum sp.	Euglena sp.
Cladophora sp.	Rhizoclonium sp.	Gloeocapsa sp.	Mallomonas sp.
Closterium sp.	Scenedesmus sp.	Gomphosphaeria sp.	Ophiocytium sp.
Coleochaete sp.	Selenastrum sp.	Merismopedia sp.	Peridinium sp.
Cosmarium sp.	Spirogyra sp.	Microcystis sp.	Phacus sp.
Eudorina sp.	Staurastrum sp.	Oscillatoria agardhii	Synura sp.
Franceia sp.	Tetracystis sp.	Oscillatoria sp.	Vaucheria sp.
Golenkinia minutissima	Tetrahedron sp.	Oscillatoria splendida	
Golenkinia sp.	Ulothrix sp.	Oscillatoria tenuis	
Gonium sp.	Zygnema sp.	Phormidium sp.	
Microspora sp.		Pseudanabaena sp. #1	
Mougeoutia sp.		Pseudanabaena sp. #2	
Oedogonium sp.		Pseudanabaena sp. #3	
Oocystis sp.		Spirulina sp.	

Table 4.3
Percentage samples in different MIB concentration ranges (Phoenix WTPs)

MIB Range (ng/L)	24th Street		Deer Valley		Val Vista		Union Hills	
	Raw	Fin.	Raw	Fin.	Raw	Fin.	Raw	Fin.
<50	89%	100%	81%	100%	100%	100%	100%	100%
<30	63%	96%	44%	92%	93%	100%	100%	100%
<20	48%	76%	33%	67%	70%	91%	100%	100%
<15	33%	52%	33%	50%	44%	83%	100%	100%
<10	22%	32%	26%	29%	33%	61%	100%	100%
<5	11%	20%	7%	8%	7%	13%	93%	96%
<2	7%	12%	0%	4%	0%	9%	86%	54%
# 2003 Samples	27	25	27	24	27	23	28	28

Table 4.4
Percentage samples in different MIB concentration ranges (Tempe & Peoria WTPs)

MIB Range (ng/L)	N. Tempe		S. Tempe		Greenway	
	Raw	Fin.	Raw	Fin.	Raw	Fin.
<50	83%	100%	100%	100%	92%	100%
<30	58%	92%	100%	100%	80%	100%
<20	42%	79%	100%	100%	52%	100%
<15	25%	67%	100%	100%	48%	100%
<10	25%	33%	88%	88%	36%	96%
<5	4%	8%	40%	40%	20%	96%
<2	4%	4%	8%	8%	12%	84%
# 2003 Samples	24	24	25	25	25	25

Table 4.5
SPME results for isolated algal cultures from the City of Phoenix water supply system

Taxonomy	*Additional Information*	*Results (ng/L)*							
		Cis-3-hexen-1-ol	*Cis-3-hexenyl acetate*	*2-isobutyl-3-methoxypyrazine (IBMP)*	*2-Isopropyl-3-methoxypyrazine (IPMP)*	*Methylisobomeol*	*β-cyclocitral*	*Geosmin*	*2,4,6-Tribromoanisole*
Pseudanabaena	*Salt River Saguaro inlet Plankton (05/01)*	*<50*	*<5*	*<2*	*<2*	*<2*	*20*	*<2*	*<2*
Phormidian	*Central Ave Canal*	*<50*	*<5*	*<2*	*<2*	*<2*	*140*	*<2*	*<2*
Anabaena	*CAP canal at cross-connect (08/01)*	*<50*	*<5*	*<2*	*<2*	*<2*	*109*	*<2*	*<2*
Pseudanabaena	*Plankton from Saguaro (08/01)*	*<50*	*<5*	*<2*	*<2*	*<2*	*134*	*<2*	*<2*
Pseudanabaena	*Plankton from Saguaro (08/01)*	*<50*	*<5*	*<2*	*<2*	*<2*	*576*	*<2*	*<2*
Plectonema *might be* **Limnothrix (Oscillatoria)**	*Plankton from Saguaro (09/01)*	*<50*	*<5*	*<2*	*<2*	*<2*	*10*	*<2*	*<2*
Pseudanabaena	*Periphyton from Saguaro (09/01)*	*<50*	*<5*	*<2*	*<2*	*<2*	*6*	*<2*	*<2*
Pseudanabaena	*From 19th Ave Canal*	*<50*	*<5*	*<2*	*<2*	*<2*	*9*	*<2*	*<2*

(continued)

Table 4.5 (Continued)
SPME results for isolated algal cultures from the City of Phoenix water supply system

Taxonomy	*Additional Information*	*Results (ng/L)*							
		Cis-3-hexen-1-ol	*Cis-3-hexenyl acetate*	*2-isobutyl-3-methoxypyrazine (IBMP)*	*2-Isopropyl-3-methoxypyrazine (IPMP)*	*Methylisobomeol*	*β-cyclocitral*	*Geosmin*	*2,4,6-Tribromoanisole*
Lyngbya	*Sediment from Saguaro (05/01)*	<50	<5	<2	<2	<2	4	<2	<2
Symploca	*Plankton from Saguaro (08/01)*	<50	<5	<2	<2	<2	15	<2	<2
Limnothrix (Oscillatoria)	*Plankton from Saguaro (09/01)*	<50	<5	<2	<2	2	83	<2	<2
Leptolyngya (Oscillatoria)	*Plankton from Saguaro (08/01)*	<50	<5	<2	<2	<2	<2	<2	<2
Microcoleus	*Plankton from Saguaro (09/01)*	<50	<5	<2	<2	<2	11	<2	<2
Limnothrix (Oscillatoria)	*Plankton from Saguaro (08/01)*	<50	<5	<2	<2	<2	45	<2	<2
Leptolyngya (Oscillatoria)	*Plankton from Saguaro (08/01)*	<50	<5	<2	<2	<2	3	<2	<2

(continued)

Table 4.5 (Continued)
SPME results for isolated algal cultures from the City of Phoenix water supply system

Taxonomy	*Additional Information*	*Results (ng/L)*							
		Cis-3-hexen-1-ol	*Cis-3-hexenyl acetate*	*2-isobutyl-3-methoxypyrazine (IBMP)*	*2-Isopropyl-3-methoxypyrazine (IPMP)*	*Methylisobomeol*	*β-cyclocitral*	*Geosmin*	*2,4,6-Tribromoanisole*
Microcoleus	*Floating pad from Saguaro (08/01)*	<50	<5	<2	<2	<2	2	<2	<2
Plankothrix (Oscillatoria)	*Periphyton from Saguaro (05/01)*	<50	<5	<2	<2	<2	19	<2	5
Pseudanabaena	*Plankton from Saguaro (09/01)*	<50	<5	<2	<2	<2	11	<2	<2
Pseudanabaena	*Plankton from Saguaro (08/01)*	<50	<5	<2	<2	<2	10	<2	<2
Lyngbya	*Plankton from Saguaro (08/01)*	285	<5	<2	<2	<2	514	<2	<2

Table 4.6
Microcystin concentrations (μg/L) in water treatment plants

Water Treatment Plant Removal—August 28, 2002			
	Deer Valley	*Squaw Peak*	*Union Hills*
Influent	*0.32*	*0.15*	*0.12*
Following PAC addition	*< 0.1*	*0.14*	*0.15*
Following chlorination	*0.11*	*NA*	*0.15*

Table 4.7
Microcystin and chlorophyll-a concentrations in three water reservoirs

	July '02	August '02	September '02	October '02
Lake Pleasant				
MC (μg/L)	0.53 (0.10)	0.12 (0.15)	0.11 (0.15)	0.16 (0.02)
Chl-a (μg/L)	0.3	0.4	0.9	1.4
Bartlett Lake				
MC (μg/L)	0.43 (0.15	0.26 (0.20)	0.26 (0.20)	<0.147
Chl-a (μg/L)	7.1	0.7	1.5	4.8
Saguaro Lake				
MC (μg/L)	0.27 (0.12	0.21 (0.10)	< 0.10	0.16 (0.00)
Chl-a (μg/L)	10.7	1.3	3.5	3.1

Note: MC concentrations are an average value (n = 3). Standard deviations are listed in parentheses.

Table 4.8
Microcystin concentrations in Lake Pleasant, Bartlett and Saguaro Lakes

	Microcystin concentration (μg-MC/L)		
	Extracellular	Intracellular	Total
a) August 19, 2002			
Lake Pleasant	< 0.20	0.000972	< 0.20
Bartlett Lake	0.26	0.000556	0.260556
Saguaro Lake	0.21	0.000506	0.210506
b) October 30,2002			
Lake Pleasant (n = 3)	0.15	> 0.005	0.15
Bartlett Lake (n = 3)	< 0.147	0.00170	< 0.147
Saguaro Lake (n = 3)	0.15	0.00175	0.15175

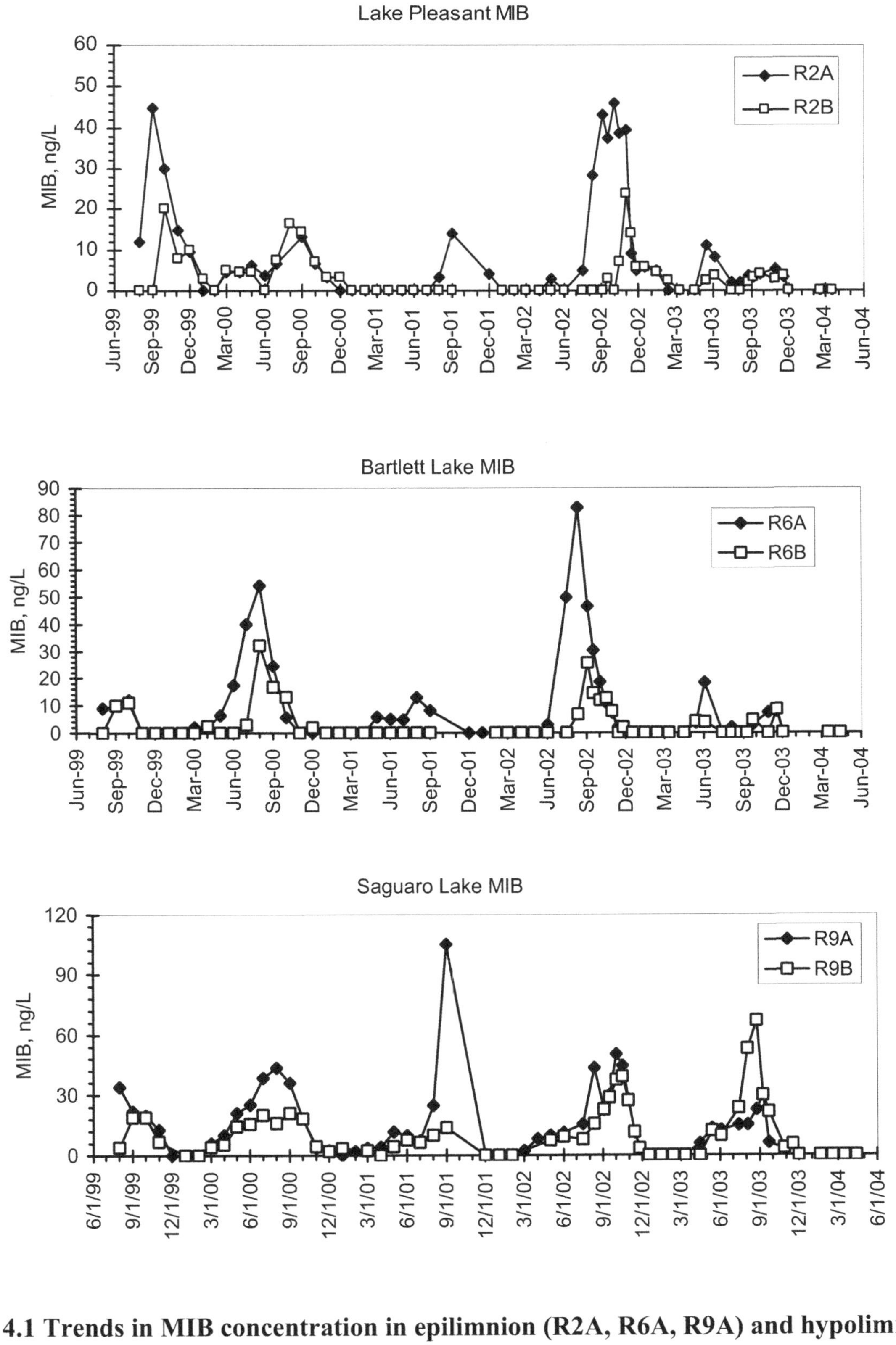

Figure 4.1 Trends in MIB concentration in epilimnion (R2A, R6A, R9A) and hypolimnion (R2B, R6B, R9B) of terminal reservoirs for the metro Phoenix region

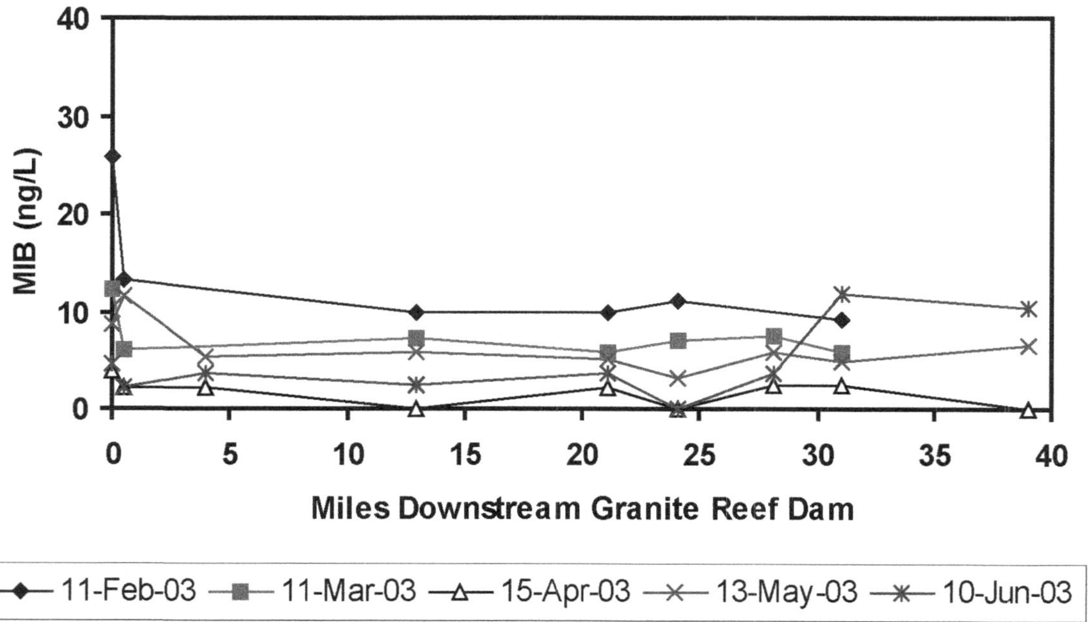

Figure 4.2 MIB Concentration along the length of Arizona Canal during cooler months

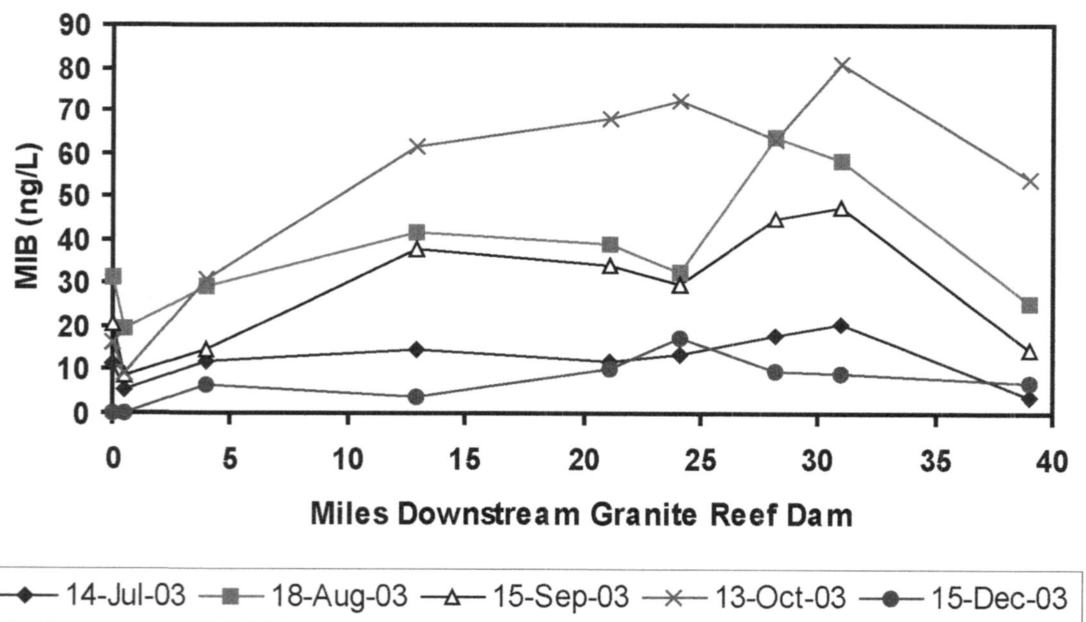

Figure 4.3 Concentrations along the length of the Arizona canal during cooler warmer water months (lower).

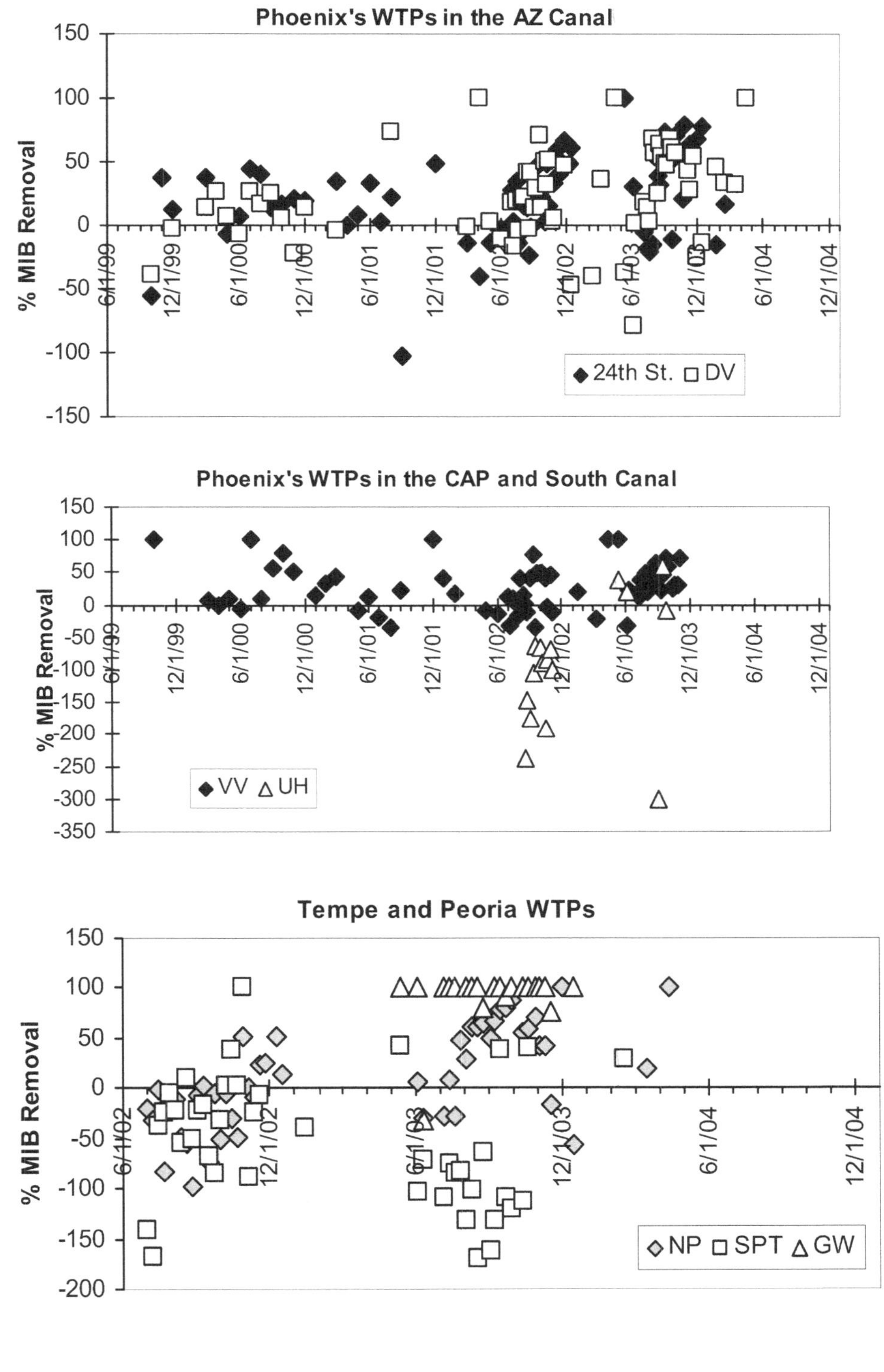

Figure 4.4 Percentage MIB removal at different WTPs

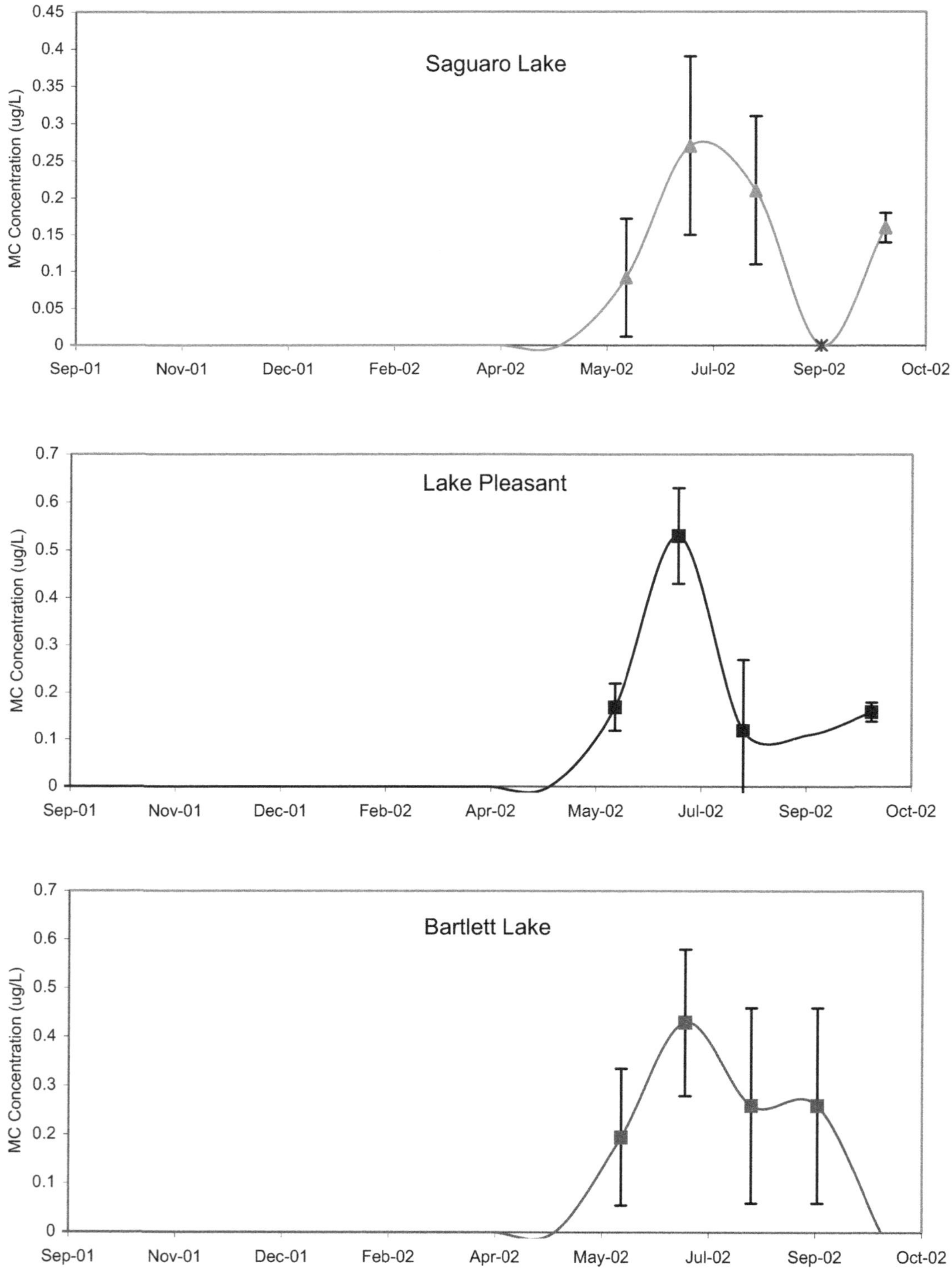

Figure 4.5 Microcystin concentrations over time in three reservoirs

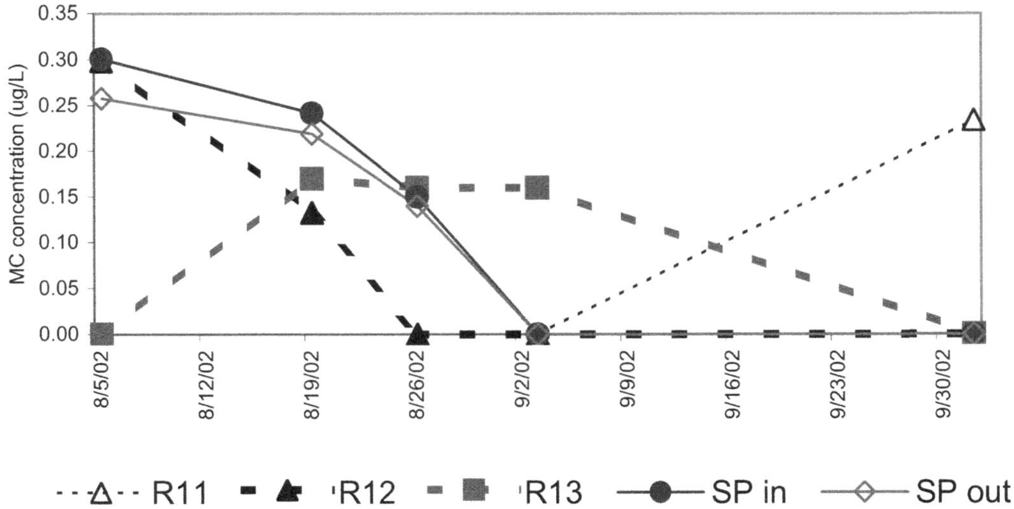

Figure 4.6 Microcystin concentrations over time in canals and 24th Street WTP. SP_{in} indicates the MC concentration entering the WTP. SP_{out} indicates the MC concentration following PAC concentration. Each data point represents the average value for three assay wells. Pooled standard deviation for all points was 0.15.

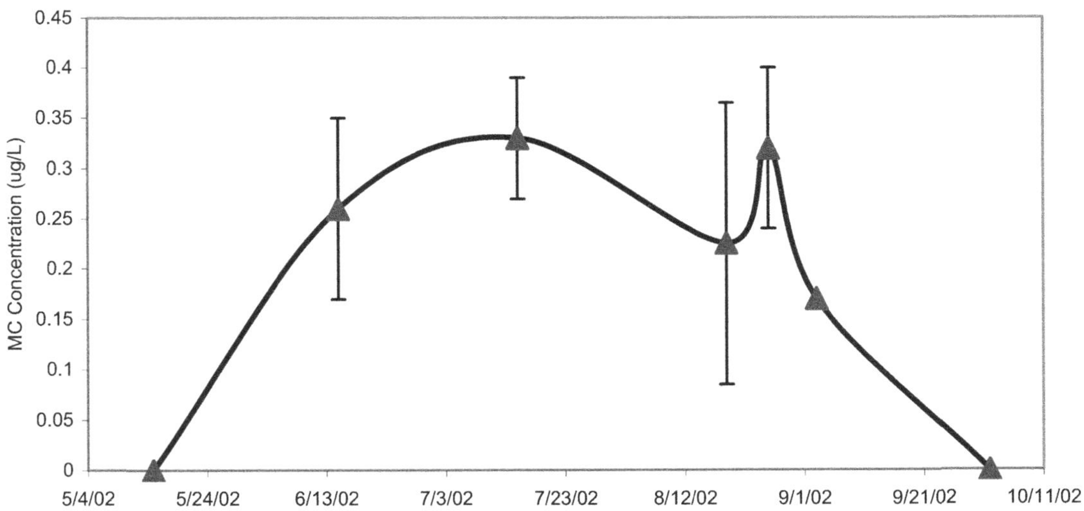

Figure 4.7 Microcystin concentration in Deer Valley WTP prior to PAC addition (n=3).

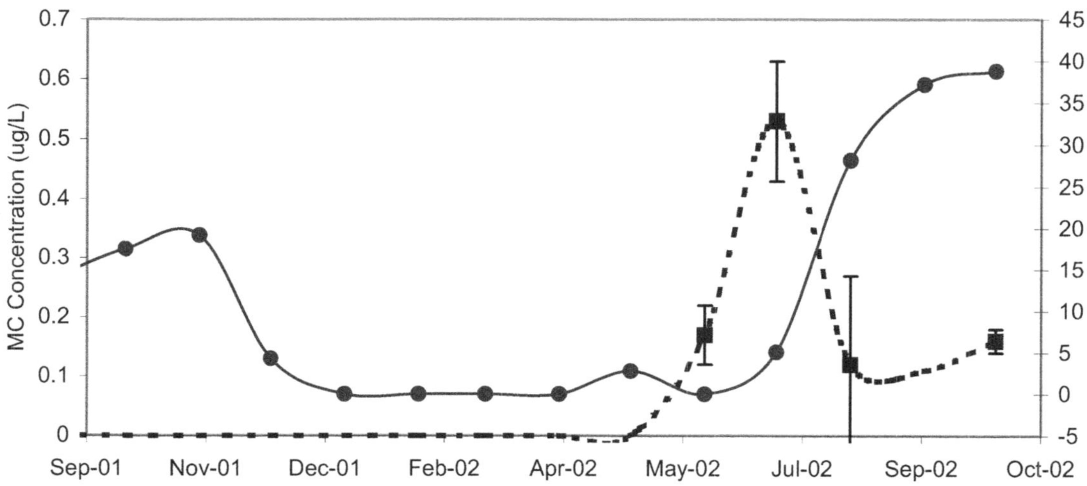

Figure 4.8 Lake Pleasant microcystin and MIB concentration with time. Microcystin is represented as the dashed line.

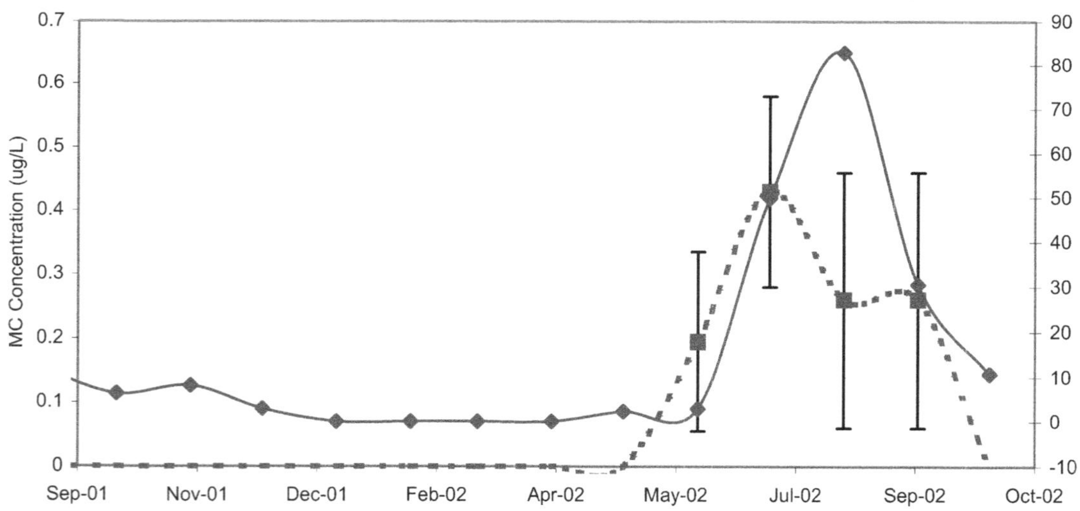

Figure 4.9 Bartlett Lake microcystin and MIB concentration with time. Microcystin is represented as the dashed line.

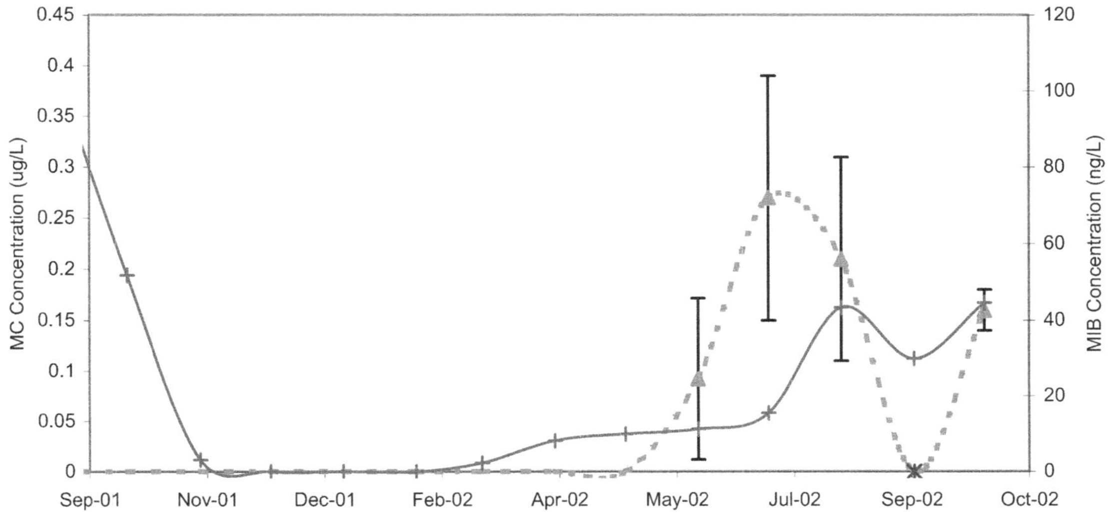

Figure 4.10 Saguaro Lake microcystin and MIB concentration with time. MC is represented as the dashed line.

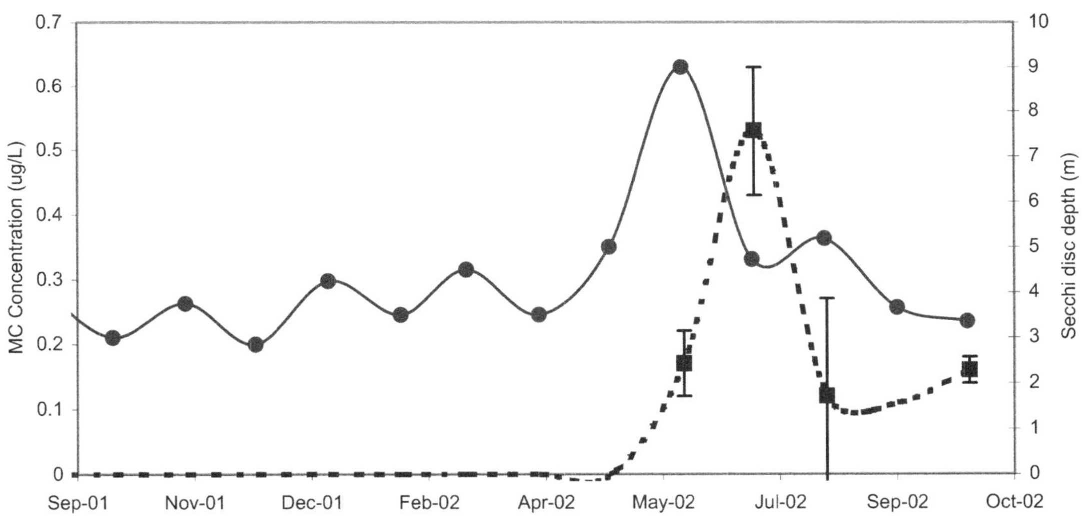

Figure 4.11 Lake Pleasant microcystin concentration and secchi disc depth with time. Microcystin is represented as a dashed line.

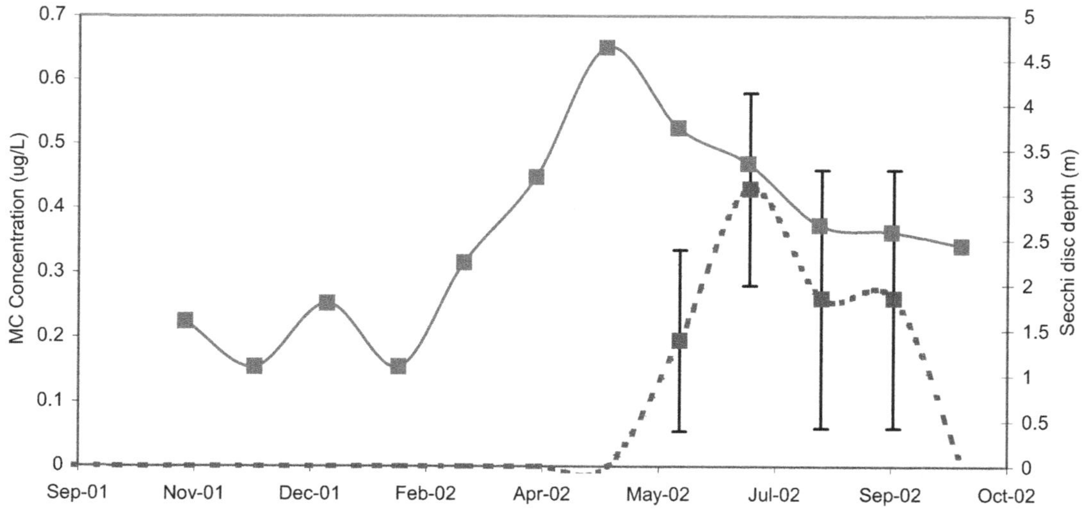

Figure 4.12 Bartlett Lake microcystin concentration and secchi disc depth with time. Microcystin represented as a dashed line.

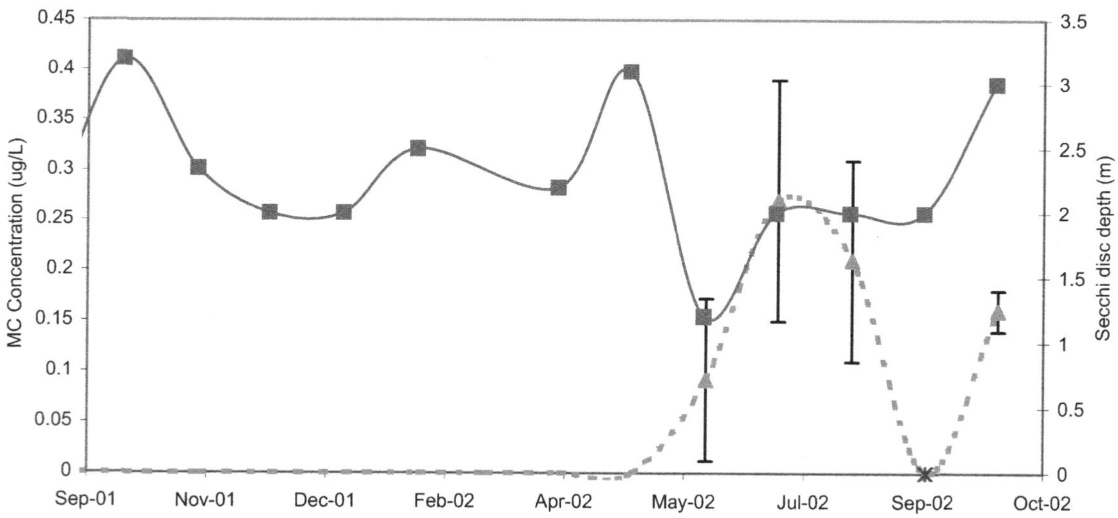

Figure 4.13 Saguaro Lake microcystin concentration and secchi disc depth with time. Microcystin represented as a dashed line.

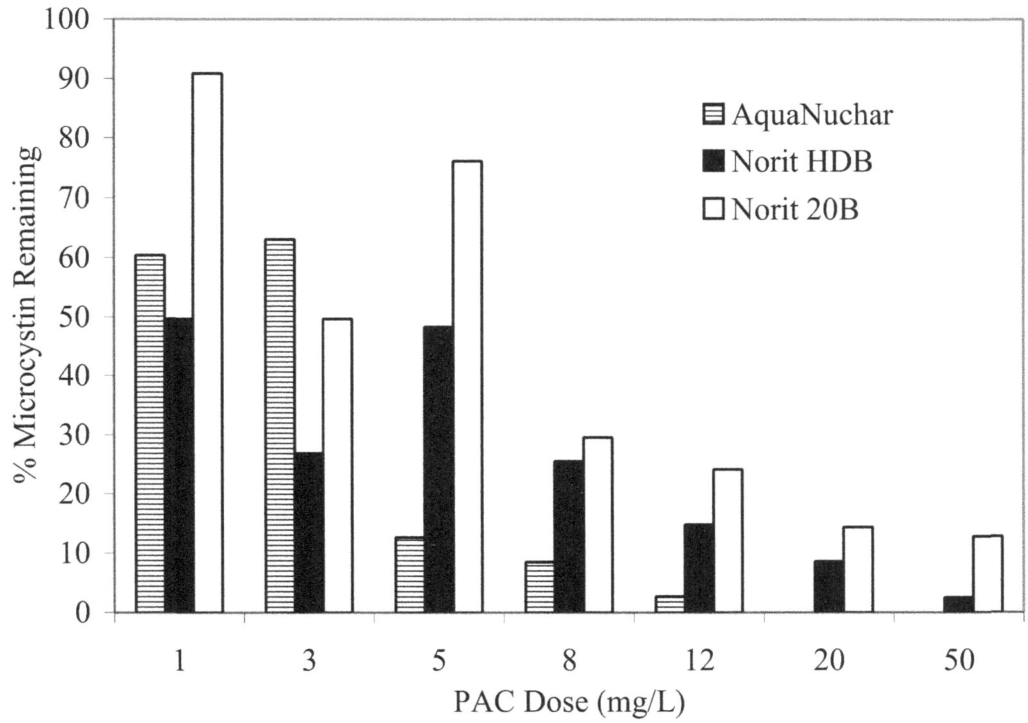

Figure 4.14 Absorption of microcystin by three powder activated carbon. Contact time of 5 hours. Initial microcystin dose of 7.5 µg-MC/L in Salt River Project water.

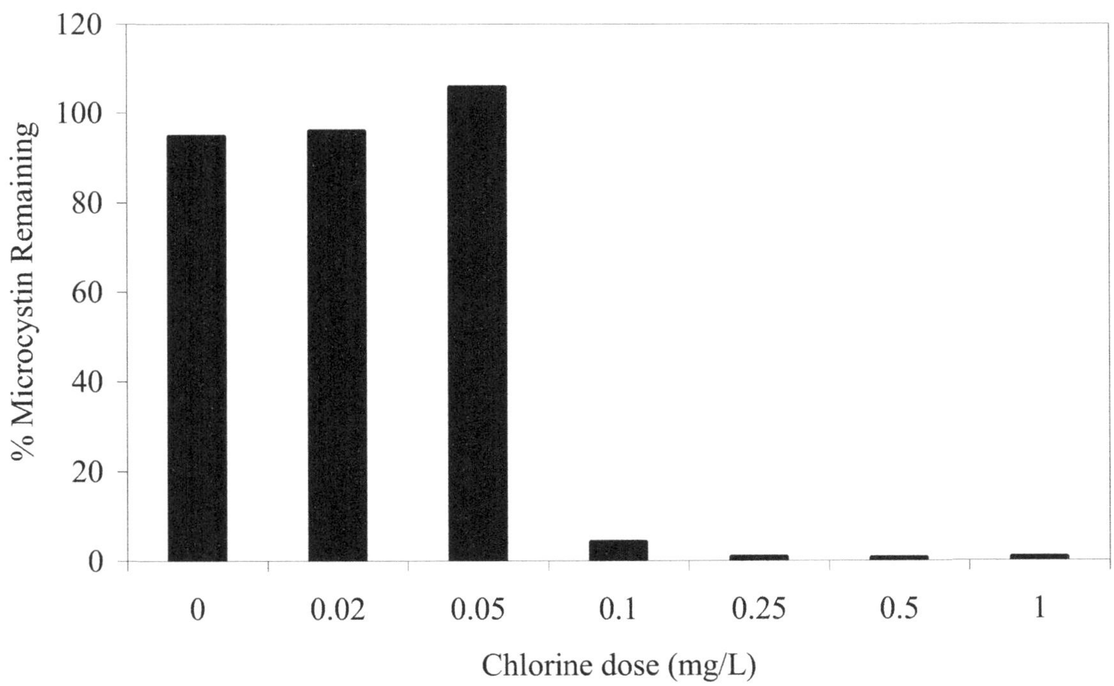

Figure 4.15 Oxidation of microcystin by chlorine after 14-day contact time

CHAPTER 5
WATER DISTRIBUTION SYSTEM MONITORING

ROUTINE SAMPLING

Twenty sampling points per month in the City of Phoenix distribution system were monitored for this study period. Sample locations were associated with each of the five water treatment plants. A total of 325 samples were collected from July 2001 to January 2003.

Sample collection occurred between the 15th and 22nd of each month. Field tests for pH, temperature and residual chlorine were performed by Pollution Control Inspectors. Samples were collected in 250 mL amber and one- liter amber bottles. Samples were analyzed for T&O by FPA and GC/MS, TDS, temperature, chlorine residual, nitrate, and bacterial enumeration.

T&O were calculated each month and graphed by date. The sum of MIB, Geosmin and β-cyclocitral were used to graph total T&O in ng/L and are illustrated in Figures 5.2 through 5.6. The Union Hills sites ranged from non detect to 16 ng/L, Val Vista sites ranged from non detect to 38 ng/L, Deer Valley sites from 2 to 36 ng/L, 24th Street sites from 2 to 51 ng/L, and Verde sites from 2 to 42 ng/L. The highest T&O months were July, August, September and October for both 2001 and 2002. A sharp decline was seen in November of each year. For example, Verde site V01 was 40 ng/L total T&O in October 2002. When the same site was sampled in November 2002 the result was 12 ng/L total T&O. FPA panel ratings for the highest T&O sampling periods identified the odors as earthy/musty, algae, grassy, and swampy. Table 5.1 lists addresses and corresponding sampling codes.

All other analyses performed for the routine baseline monitoring fell within the "Typical Water Analysis" ranges for the City of Phoenix distribution system. Ranges observed were as follows: Temperature 9°C – 41°C, pH 7.2 – 8.48, TDS 328 – 702 mg/L, Nitrate < 0.1 – 1.6 mg/L, and HPC 7 day < 20 – 3000 CFU/mL.

RESPONSIVE DISTRIBUTION SYSTEM SAMPLING

Complaint Inquiry forms (Appendix A) have been developed to respond to Water Services Department customer complaints. The Pollution Control Division inspectors asked the customer a series of questions before performing a site visit. Inspectors encouraged the customer to describe the color/smell or taste to determine the possible cause. In addition, the inspector documented when the problem occurred and for how long. Other questions asked include those dealing with water softener use and service and whether construction or repairs have been made near the home. In addition, the customer may be asked to by-pass the softener to see if the water runs clear.

When performing a site visit the inspectors sample from an outside tap. After flushing for five minutes samples are taken for T&O and bacti. Field chlorine tests are performed on site.

No significant customer complaints were received during the study period. As discussed in chapter 2 the City of Phoenix maintained a 2.5 mgCl_2/L residual leaving the water treatment plants for the entire distribution system sampling effort.

Baseline distribution system monitoring did not show significant differences from MIB or Geosmin leaving the water treatment plants. Therefore, at the selected sampling points distribution system production of T&O did not occur or could not be detected.

INTENSIVE TEMPORAL SAMPLING CAMPAIGNS

The City of Phoenix obtained samples from a historically problematic area of the distribution system (Figure 5.1). The sampling effort included public buildings (Fire and Police stations, library, hospital, school) water fountains, and private homes. Thirty samples were collected on the same day south and west of the City of Phoenix Deer Valley Water Treatment Plant. The sample effort was conducted from early morning throughout the day. Samples were analyzed for by FPA, GC/MS, TDS, pH, temperature, nitrate, chlorine residual, and actinomycete and HPC (7 days).

Samples were collected from the distribution sites around the Deer Valley WTP on October 1, 2003. At the time of sampling, assumptions were made that all the water for these sampling sites originated from the Deer Valley WTP. Information provided by SCADA operators revealed several discrepancies. Three of the sites were in pressure zones 3A or 3D which was not supplied by the Deer Valley WTP. There were several occasions shown on the SCADA report where there was flow from the Deer Valley Reservoir # 2, but no pumps were running. This indicated that Zone 2A was supplied by the Union Hills Plant or by wells. In addition, fluctuations in flow rates made it difficult to predict travel times from the Deer Valley WTP. Fifteen sites were positively identified as those with water from the Deer Valley WTP. Each site was ranked based on distance from the plant. Table 5.2 summarizes the GC/MS and FPA data for the 15 ranked sites.

Twenty of the thirty samples collected were analyzed by the FPA panels and by GC/MS. Of the twenty sites, six were ranked as having objectionable descriptors by the panel. The closest sample to the plant had relatively low concentrations of T&O compounds with MIB at 4 ng/L and Geosmin at 5 ng/L. However, the FPA sum for swampy was a 4 (weak). The sample farthest from the plant also had relatively low concentrations of T&O compounds with MIB at 8 ng/L and Geosmin at 6 ng/L and the FPA sum for swampy was 2.5 (very weak). In addition, this site had high HPC results. Two additional sites also had relatively low concentrations of MIB and Geosmin and odors identified as swampy. Two sites also rated as objectionable had higher MIB and Geosmin concentrations. One site rated as Earthy/Musty by the panel had MIB at 31 ng/L and Geosmin at 7 ng/L. The other site which had MIB at 12 ng/L, Geosmin at 5 ng/L and β-cyclocitral at 4 ng/L was also rated as Earthy/Musty by the panel.

All other analyses performed for the monitoring fell within the "Typical Water Analysis" ranges for the City of Phoenix distribution system. Ranges observed were as follows: Temperature 9° C – 41° C, pH 7.2 – 8.48, TDS 600 – 672 mg/L, Nitrate 0.1 – 0.3 mg/L, and HPC 28 C 7 day 2 – >5700 CFU/mL. All Actinomycetes analyses were < 1 CFU/mL.

Intensive monitoring did not show significant differences in MIB or Geosmin. The data does not indicate production of tastes and odors based on time and distance from the plant. The results indicate site specific T&O issues.

CORRELATING FPA AND GC/MS DATA FOR EARTHY-MUSTY ODORS

ASU collected influent and effluent samples from four of COP's WTPs (24th Street (24th), Deer Valley (DV), Union Hills (UH) and Val Vista (VV) WTPs) on the same day and at the same locations for both FPA and GC-MS analyses. A total of 140 samples were collected from May of 2003 to March of 2004 for FPA and GC/MS analysis. Approximately 36 %, 23 %, 22 % and 19 % of the 140 samples were from UH, DV, 24th, and VV WTPs. FPA samples were collected in 1-L odor-free amber bottles provided by COP's WTPs. Samples were delivered to two FPA panels at DV and UH WTPs within the same day of collection and analyzed within 24 hours. ASU provided COP with the sampling schedule at least three months in advance, so that FPA panels can be coordinated accordingly. GC-MS samples were collected in precleaned 125 mL MIB amber bottles (acid washed in 10 M HCl for 12 hours and ashed at 550° C for 2 hours).

For analytical purposes, all GC/MS results with values below the detection limit (< 2.0 ng/L for MIB and Geosmin and < 5.0 for β-cyclocitral) were given a value of 1 ng/L. Weber-Fechner plots for MIB, Geosmin and β-cyclocitral were done to determine the possible correlation between these T&O compounds and estimated/measured Flavor Profile Analysis (E/M FPA) intensity. It has been determined that changes in human perception to odor intensity are logarithmic because as odor intensity increases it requires a greater change in odor to sense the change (Meng et al. 1992).

Multiple-variable non-linear regressions (MVNLR) between the 140 paired FPA/GC-MS datapoints were also performed to determine an alternative correlation to the Weber-Fechner plots. The following four equations were used to perform MVNLRs:

$$E/M\ FPA = a*MIB + b*Geosmin + c*\beta\text{-cyclocitral} + d \qquad \text{Equation (5.1)}$$
$$E/M\ FPA = a*MIB + b*Geosmin + c*\beta\text{-cyclocitral} \qquad \text{Equation (5.2)}$$
$$E/M\ FPA = a*MIB^b + c*Geosmin^d + e*\beta\text{-cyclocitral}^f \qquad \text{Equation (5.3)}$$
$$E/M\ FPA = a*MIB^b *Geosmin^c *\beta\text{-cyclocitral}^d \qquad \text{Equation (5.4)}$$

During the first round of MVNLRs, the 140 paired datapoints were used to identify outliers that may bias the regressions. Two cases were tested considering only two (MIB and Geosmin) and all three T&O compounds. Several outliers were identified during the first round of regressions and eliminated for the second round. During the second round of regressions, 43 datapoints < 1.0 and 7 outlier datapoints between 1.5 and 5.0 were eliminated. The second round of MVNLRs gave better results than the first round ($R^2 > 0.55$). For brevity, only the best fits during the second round of regression are presented herein.

Earthy/Musty odors were not detected by the panelist in approximately 31 % of the 140 samples (No E/M FPA intensity entered in table) (Figure 5.7). Only 11 % of the samples had objectionable E/M FPA intensities (> 3). The number of samples with a certain E/M FPA Intensity decreased as the E/M Intensity increased (Figure 5.7). This inverse correlation made difficult the statistical analysis because the high variability of E/M FPA intensity < 1.

Figure 5.8 shows the Weber-Fechner plots for MIB, Geosmin and β-cyclocitral. E/M FPA intensity showed a significant direct correlation with MIB and Geosmin ($R^2 = 0.86$ and $R^2 = 0.71$, respectively), but a weak correlation with β-cyclocitral ($R^2 = 0.08$). The only problem with these plots is that the GC/MS measurements had high variability for any given E/M FPA (Figure 5.8). A logarithmic correlation was expected using the Weber-Fechner plots (Weber, 1982), however, the equation that best fitted the datapoints was a linear equation.

Figure 5.9 shows the best MVNLRs determined during the second round of regressions. E/M FPA values between 1.25 and 4 were successfully predicted with all four equations. The best equations based on R^2 were Equation 5.1 and 5.4 ($R^2 \sim 0.75$). It is important to notice that Geosmin had a negative effect on the predicted E/M FPA in equations 5.1, 5.2, and 5.4 (the higher Geosmin, the lower the predicted E/M FPA). The causes of this trend have not been determined.

E/M FPA intensity was successfully predicted with the MIB and Geosmin Weber-Fechner plots and a three component (MIB, Geosmin, β-cyclocitral) MVNLRs. However, plots and regressions may lack significance as variability was high and many outliers were eliminated during the analysis. High variability may be attributed to the fact that approximately one third of the samples originated from UH WTP, which is located on the CAP Canal, a canal that does not present T&O problems during summer and has low T&O levels throughout the year. A better reproducibility of FPA results may be obtained if more samples with higher T&O are used during FPA

SUMMARY

Baseline water distribution system monitoring indicated:

- Earthy/Musty FPA showed a direct correlation with MIB and geosmin but a weak correlation with β-cyclocitral.
- Earthy/Musty FPA intensity could be predicted with the MIB and geosmin plots and with a three component (MIB, Geosmin, β-cyclocitral) plot.
- Varibility was high for all.
- Distribution system production of T&O did not occur or could not be detected.

Table 5.1
Addresses and corresponding sampling codes

Sampling code	Address
DV01	35th Avenue/Campbell
DVO2	43rd Avenue/Northern
DV03	35th Avenue/Bethany Home
DV04	91st Avenue/Indian School
VV01	Rio Salado/Hardy
VV02	19th Avenue/Southern
VV03	Warner/Equestrian Trail
VV04	91st Avenue WWTP
SP01	24th Street/ Campbell
SP02	18th Street/Buckeye
SP03	7th Street/Camelback
SP04	3rd Avenue/McDowell
UH01	7th Street/Deer Valley
UH02	19th Avenue/Deer Valley
UH03	19th Avenue/Beardsley
UH04	19th Avenue/Grovers
V01	Highway 87/Gilbert
V02	52nd Street/Thomas
V03	54th Street/Lafayette
V04	40th Street/Indian School

Table 5.2
Taste and Odor Summary of Intensive Sampling Campaign

Address	Time Rank	MIB ng/L	Geosmin ng/L	β-cyclocitral ng/L	T&O Sum ng/L	Earthy /Musty FPA	Grassy FPA	Swampy FPA	Other FPA	FPA Sum
9401 N 35th Ave.	1	4	5	0	9	0	0	4		4
8001 N. 36th Ave.	2	8	6	0	14	1.25	1	0		2.25
3120 W. Cholla Ave.	3	3	4	6	13	1.25	1	0		2.25
8511 N. 17th Ave.	4	12	5	4	21	3.5	0	0		3.5
8021 N. 43rd Ave.	5	3	4	7	14	1	1	1.25	Sulfur 0.5	3.25
7001 N. 29th Ave.	6	5	7	0	12	0	0	1.25	Sulfur 1	1.25
8026 N. 16th Ave.	7	8	6	0	14	1.75	2	0	Sulfur 0.75	3.75
10619 N. 43rd Ave.	8	8	5	4	17	1.5	0	1		2.5
6142 N. 35th Ave.	9	34	8	5	47	1	0	1		2
6014 N. 30th Ave.	10	8	6	0	14	0	0	2.5		2.5
7901 N. Central	11	12	0	0	12	1.5	0	0		1.5
403 E. Hatcher Rd.	12	8	04	0	12	1.5	0.75	0		2.25
2000 W. Bethany Home Rd.	13	9	03	0	12	3	0	0		3
6002 N. 7th Ave.	14	16	2	0	18	1.5	0	0.5		2
7036 N. 7th St.	15	17	0	0	17	1	2	2		5

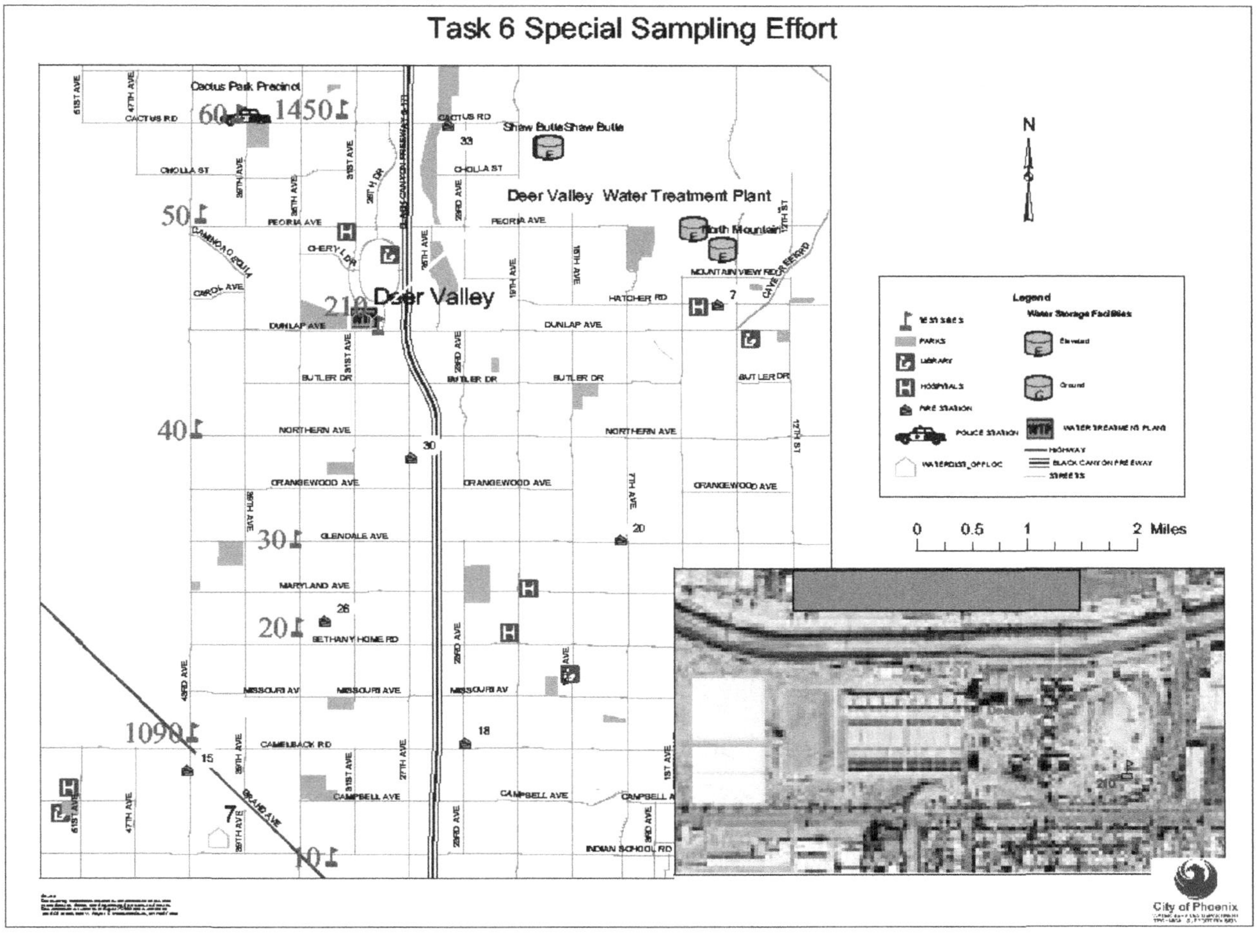

Figure 5.1 Schematic of City of Phoenix distribution system sampling.

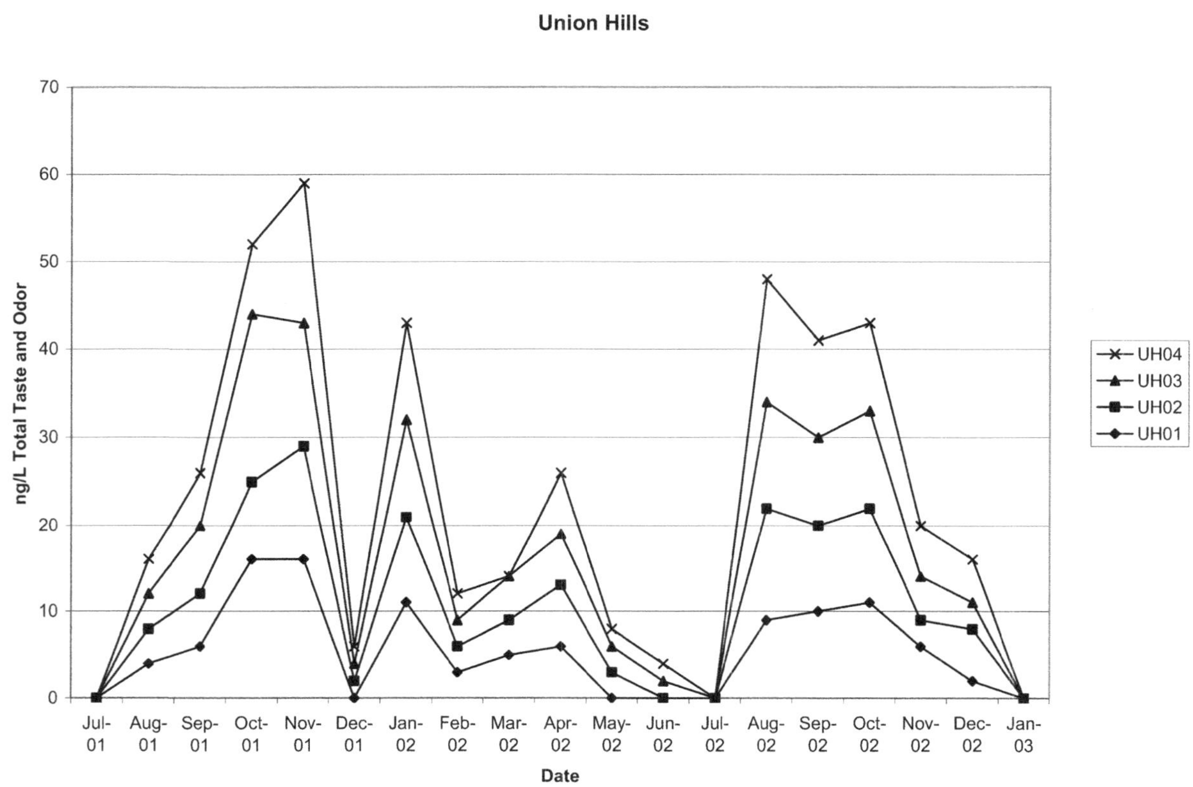

Figure 5.2 Monitoring results for Union Hills Water Treatment Plant.

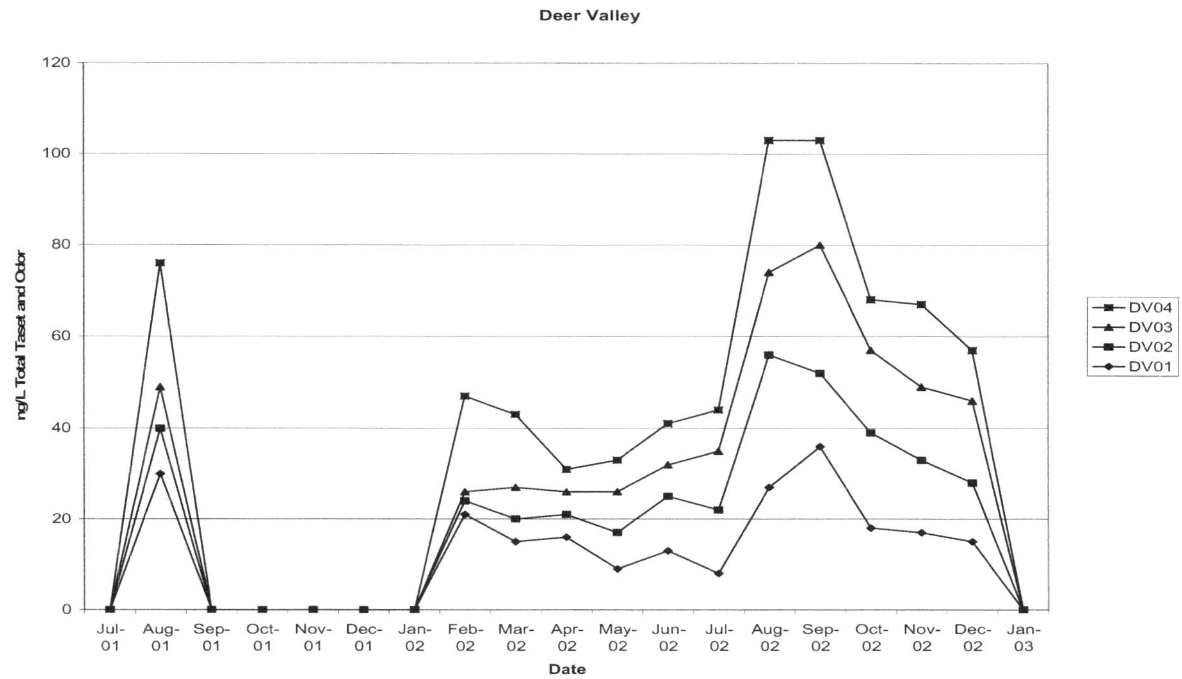

Figure 5.3 Monitoring results for Deer Valley Water Treatment Plant

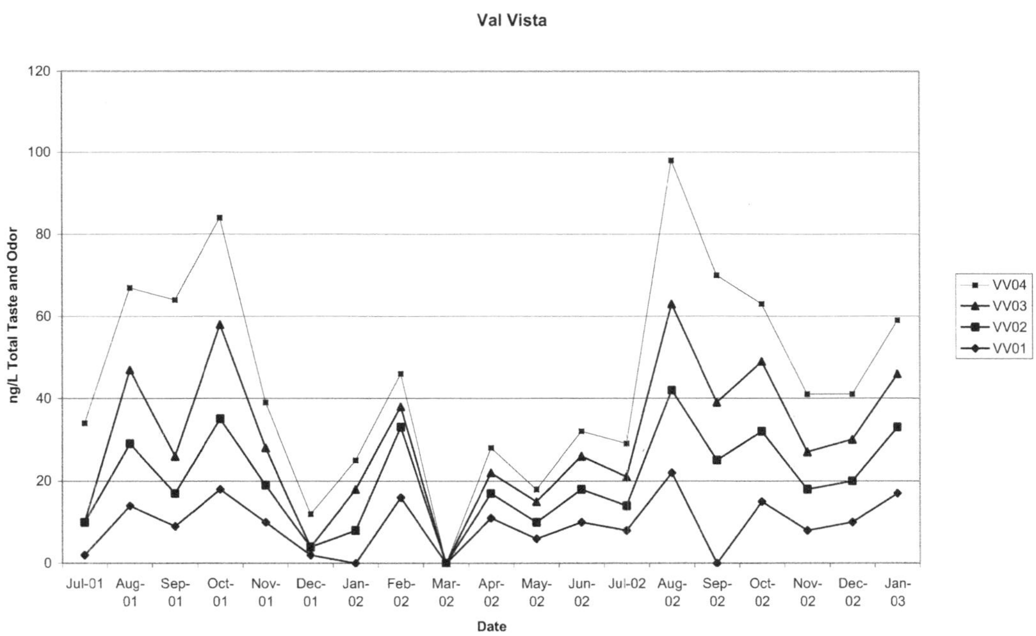

Figure 5.4 Monitoring results for Val Vista Water Treatment Plant

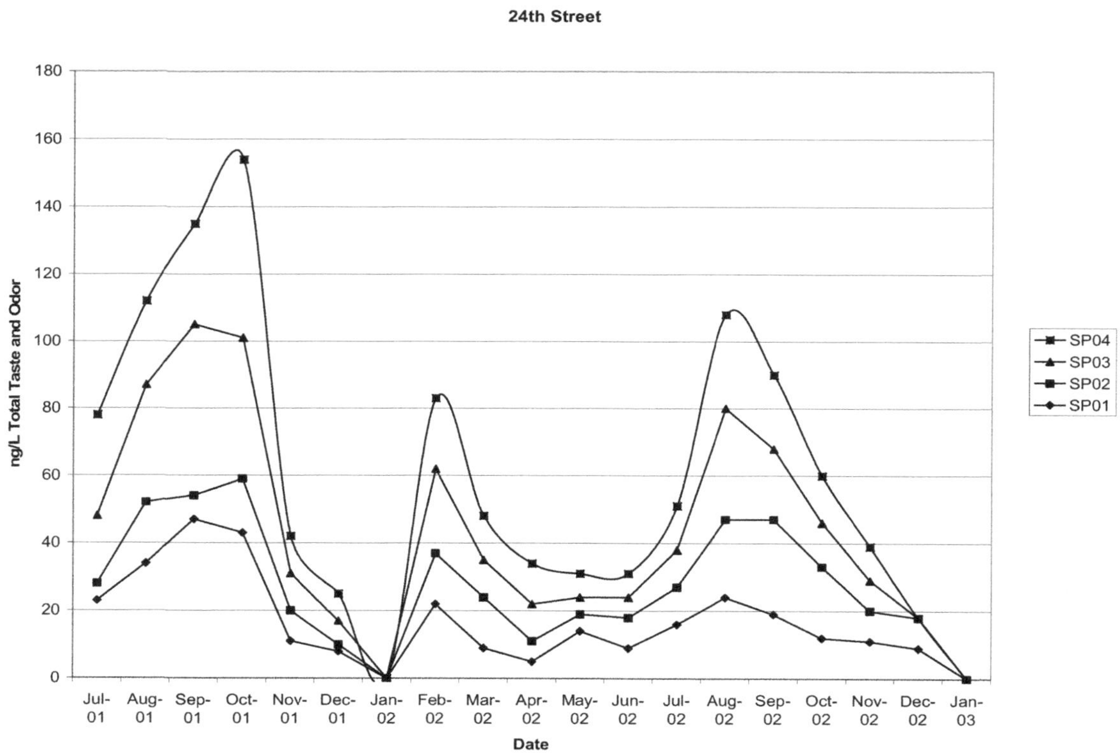

Figure 5.5 Monitoring results for 24th Street Water Treatment Plant

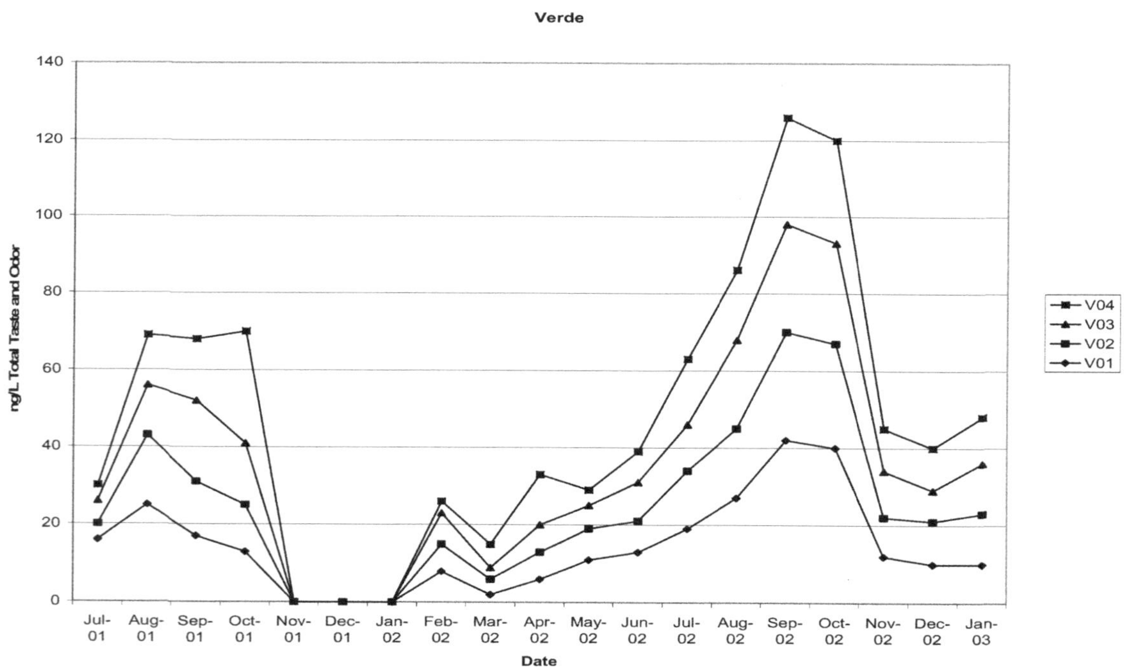

Figure 5.6 Monitoring results for Verde Water Treatment Plant

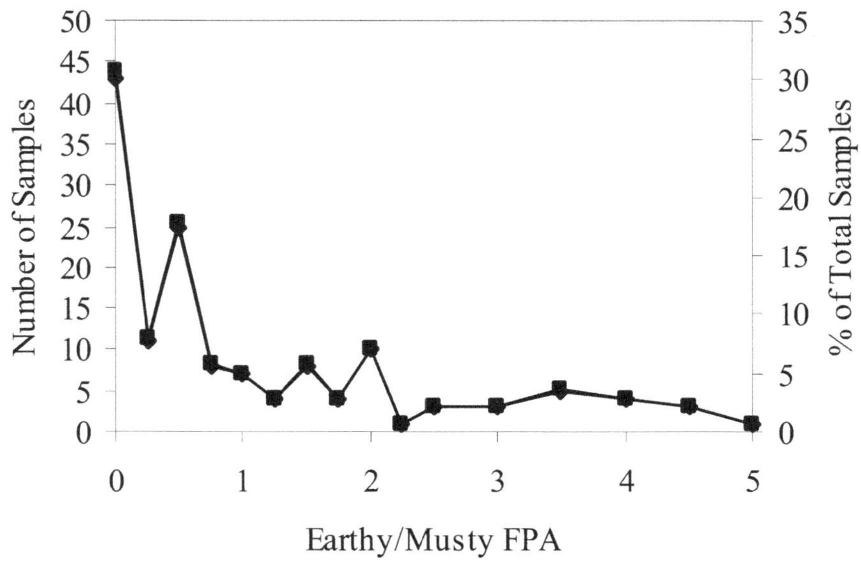

Figure 5.7 Number of samples vs. Earthy/Musty FPA.

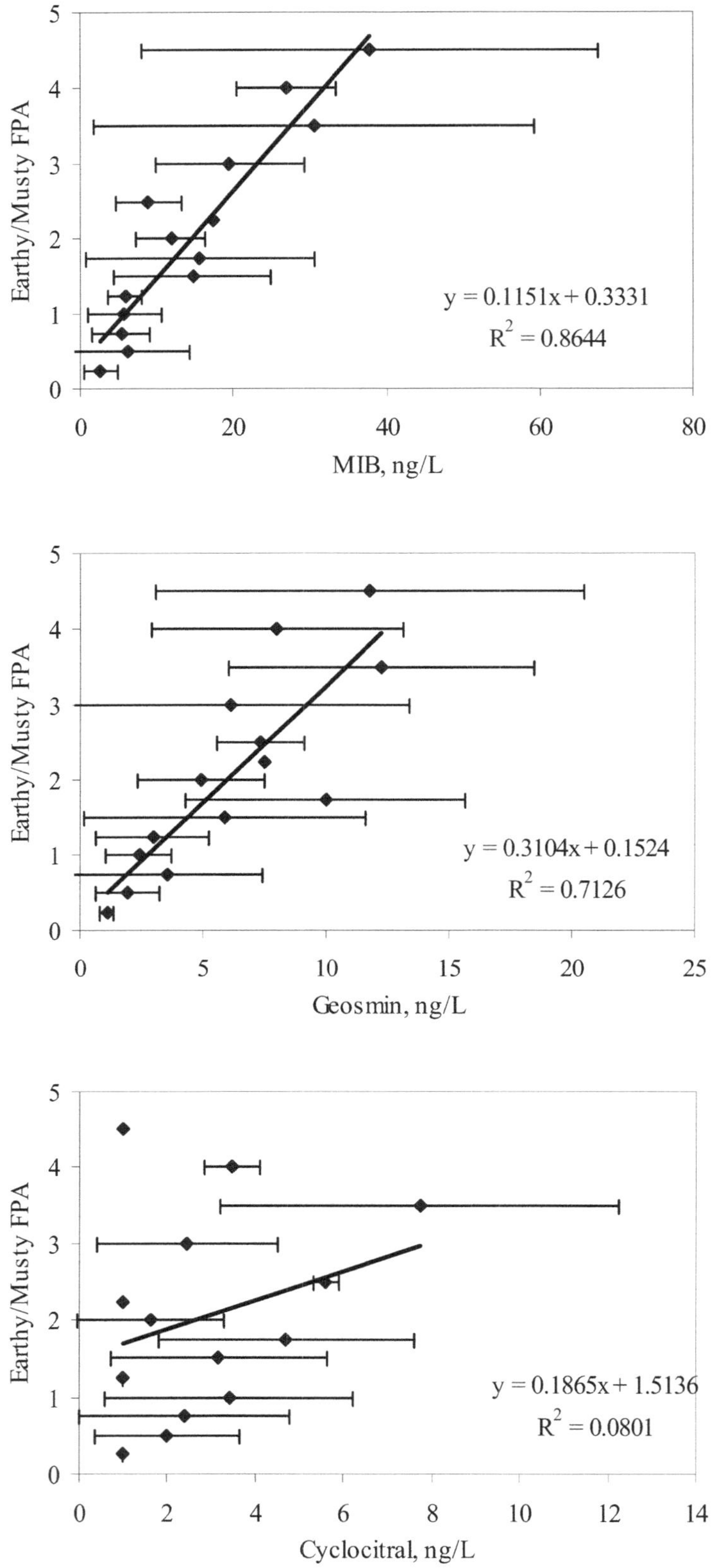

Figure 5.8 Multiple-Variable Non-Linear Regressions between Earthy/Musty FPA and three T&O Compounds.

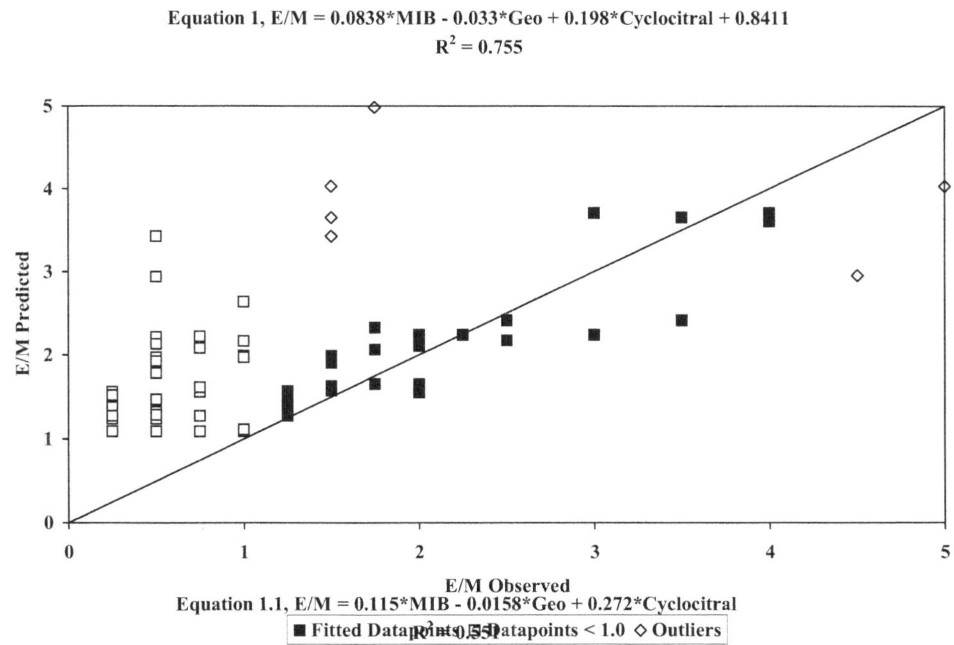

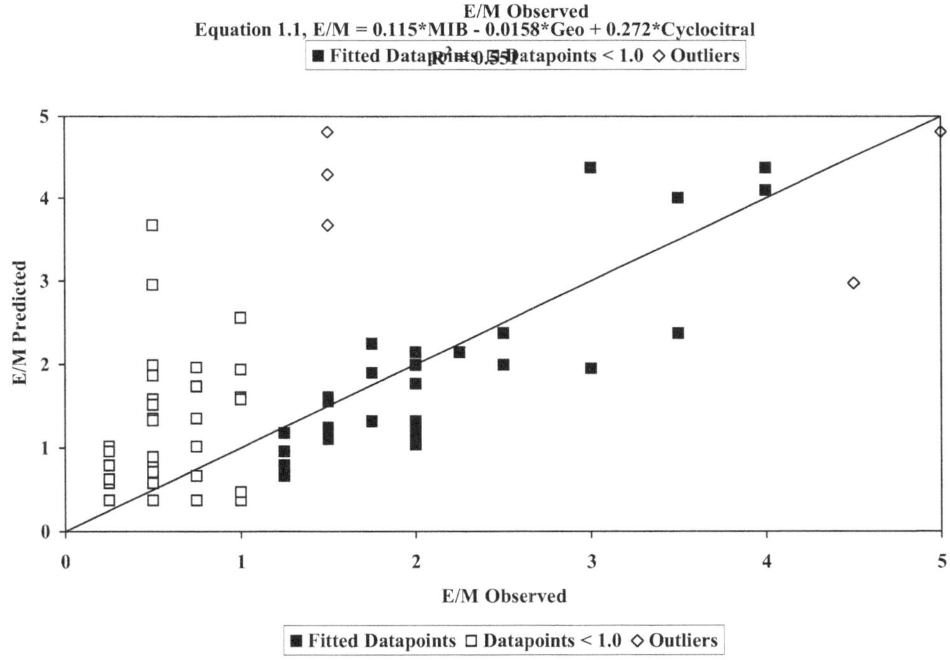

Figure 5.9 Earthy/Musty FPA predicted with multiple variable regression equations.

(continued)

Equation 2, $E/M = 0.2023*MIB^{0.7443} + 0.4629*Geo^{-0.2513} + 0.2747*Cyclocitral^{0.8747}$

$R^2 = 0.754$

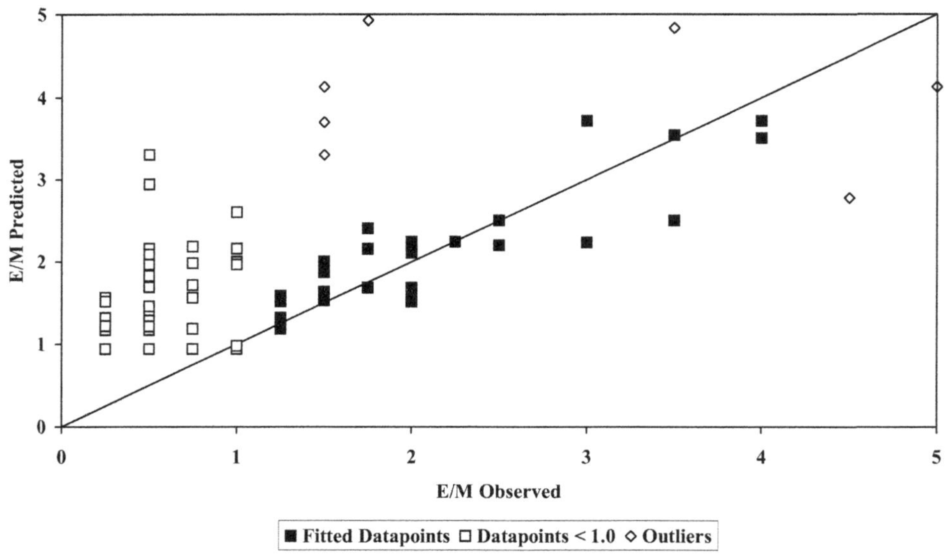

Equation 3, $E/M = 0.800*MIB^{0.396}*Geo^{-0.110}*Cyclocitral^{0.350}$

$R^2 = 0.728$

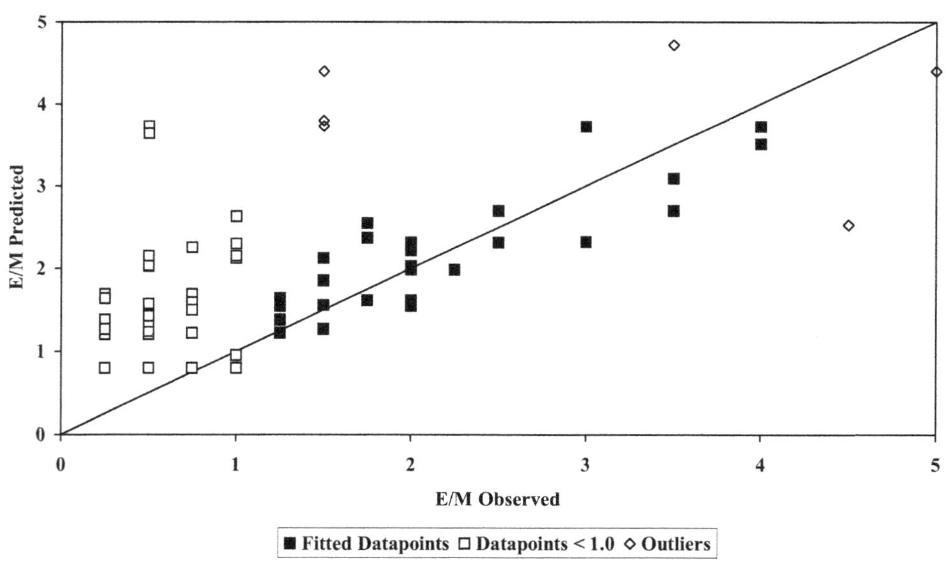

Figure 5.9 (Continued). Earthy/Musty FPA predicted with multiple variable regression equations.

CHAPTER 6
LABORATORY STUDIES OF ACTINOMYCETE INACTIVATION

This chapter presents results of laboratory tests to understand inactivation of actinomycetes, which may cause T&O production in distribution system biofilms. Batch experiments with actinomycete in the absence and presence disinfectants were conducted. With a constant disinfectant residual the Hom model can be used:

$$Ln\left(\frac{N}{N_0}\right) = -kC^n T^m \qquad (6.1)$$

Where N/N_0 is the survival ratio, k is a pseudo first-order rate constant, n and m are empirical coefficients, C is the disinfectant concentration, and T is contact time.

Two solutions were considered in the inactivation tests: 1) a peptone water solution, and 2) a phosphate buffer solution (PBS). Figure 6.1 shows the chlorine demand in the two solutions over the course of 60 minutes. A higher chlorine demand was observed in the 0.15 peptone solution, whereas the chlorine residual was nearly constant in PBS. *Streptomycete* die-off in PBS over 5 days was quite slow (Figure 6.2). Therefore, all chlorination experiments were conducted in PBS containing *Streptomycetes*, which had been previously centrifuged and re-suspended.

The effect of pH during chlorination on *Streptomycetes* inactivation in the presence of free chlorine was investigated (Figures 6.3 through 6.5). Chlorine depletion was also monitored in each experiment. The inactivation plots represent the survival ratio (N/N_0) over the 60 minute tests. N_o is the initial *Streptomycetes* concentration, and ranged from 10^9 to 10^{14} CFU/100 mL. While these values are high, the solution was not visibly highly turbid. Values for k, m and n were estimated using Mathematica version 4.2 (Wolfram, 2002). This computer program analyzed data using nonlinear data fitting procedures. Table 6.1 summarizes the Hom model parameters. For these free chlorine experiments when chlorine depletion varied over time a modified Hom model was used (Hass *et al.* 2001) which uses a second order rate constant for chlorine depletion (Table 6.1).

Streptomycetes inactivation at pH 5 occurred immediately after chlorine addition. Two to three log inactivation was observed within 10 minutes in the presence ~ 2 mg/L of free chlorine. At pH 7 and 9, the same level of inactivation was achieved within 10 to 20 minutes. CT inactivation values are summarized in Table 6.2. While some variability in CT values is observed, the lowest CT values were obtained at pH 5 and suggests that hypochlorous acid is a better disinfectant than hypochlorite ion.

Streptomycetes inactivation at pH 8.5 with monochloramine is presented in Figure 6.6. The rate of inactivation is slightly slower than free chlorine (Figure 6.5) at a similar pH.

Table 6.1
Hom model coefficients for *Strepomycete* inaction studies

Disinfectant	**pH**	**Second order rate constant for chlorine depletion (L/mg*minutes)**	***k***	m	n
Free Chlorine	5	0.0037	**0.010**	**0.58**	**5.9**
Free Chlorine	7	0.003	**0.059**	**0.78**	**2.0**
Free Chlorine	9	0.0053	**0.055**	**0.80**	**2.1**
Monochloramine	8.5	--	**0.019**	**1.1**	**2.5**

Table 6.2
CT values (mg-min/L) for *Streptomycete* at pH 5, pH 7 and pH 9 with free chlorine

pH	CT values at different percentage Inactivation		
	50%	***99%***	***99.9%***
5	4.3	**27**	**27**
7	12	**29**	**26**
9	12	**18**	**19**

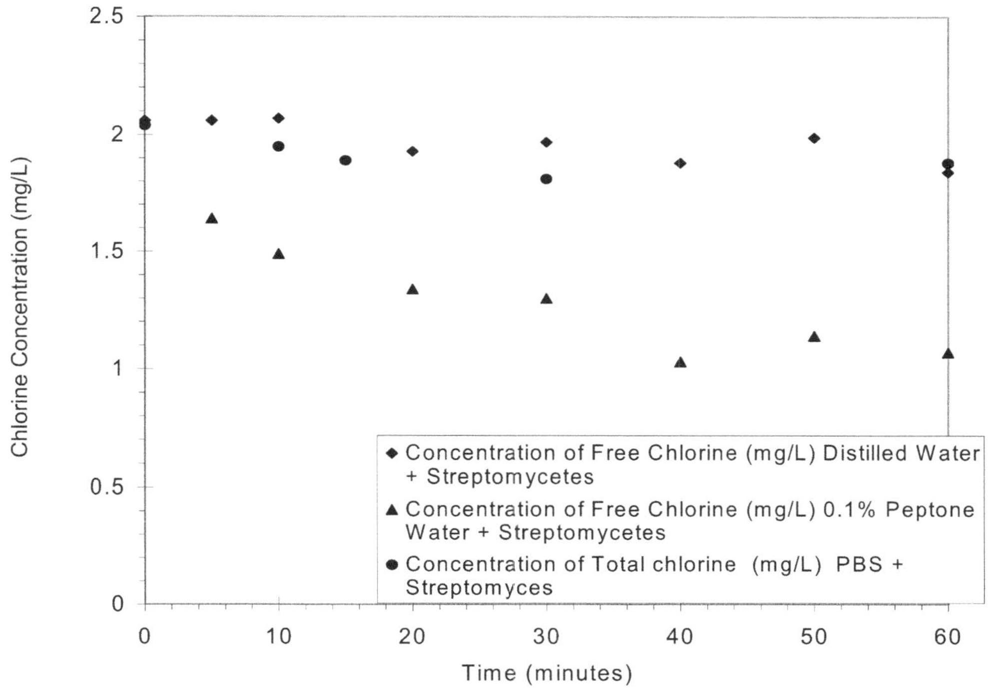

Figure 6.1 Free chlorine depletion in different water matrices conducted in triplicate experiments

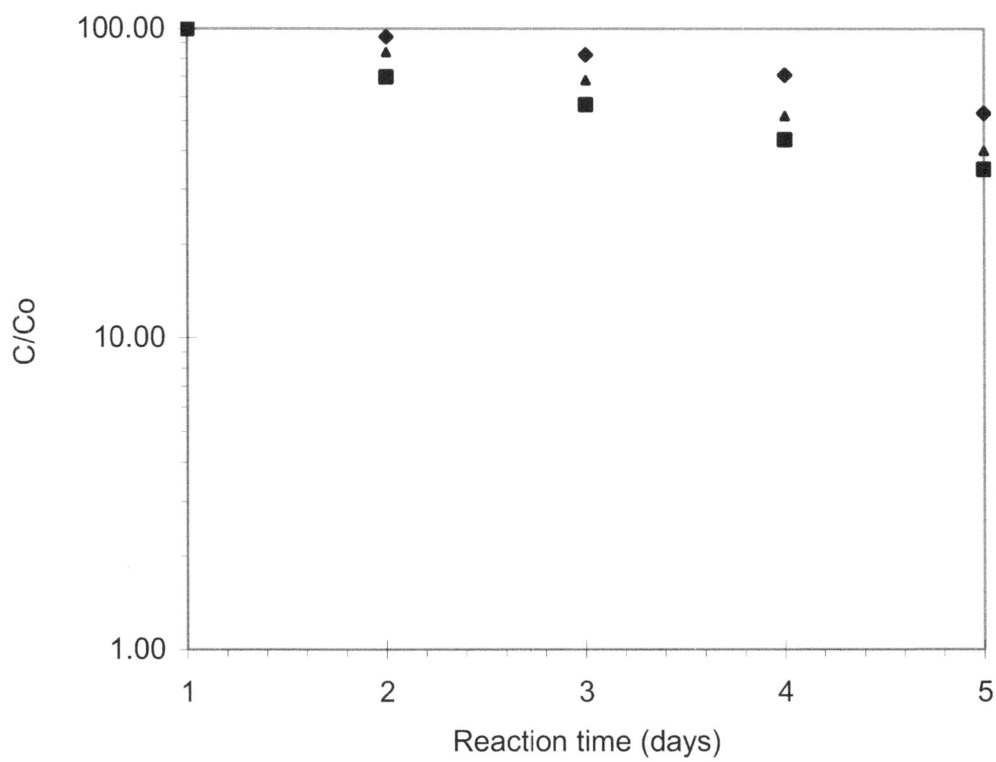

Figure 6.2 *Streptomycetes* dieoff in PBS in triplicate experiments

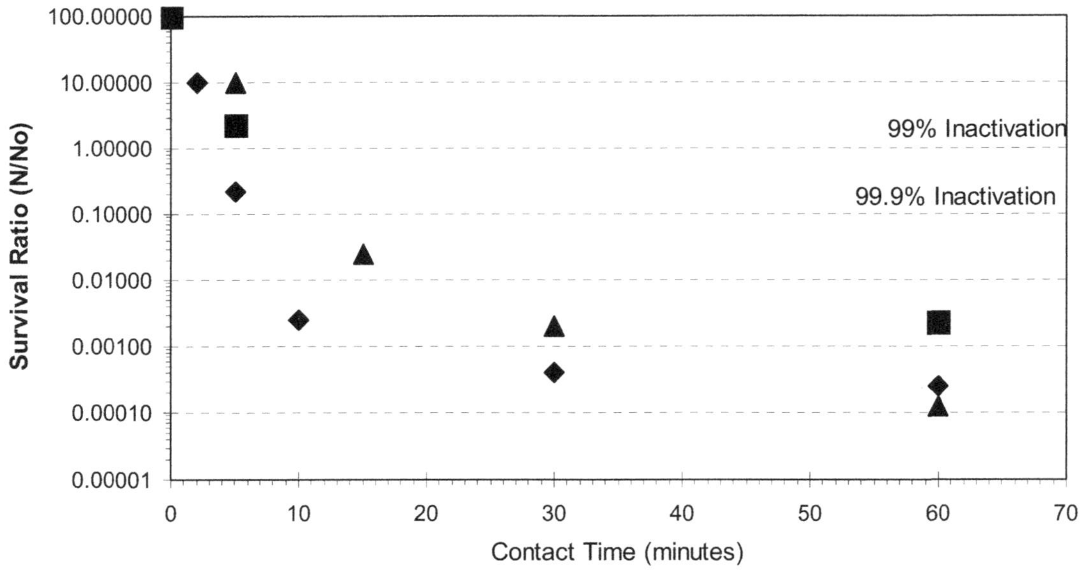

(A)

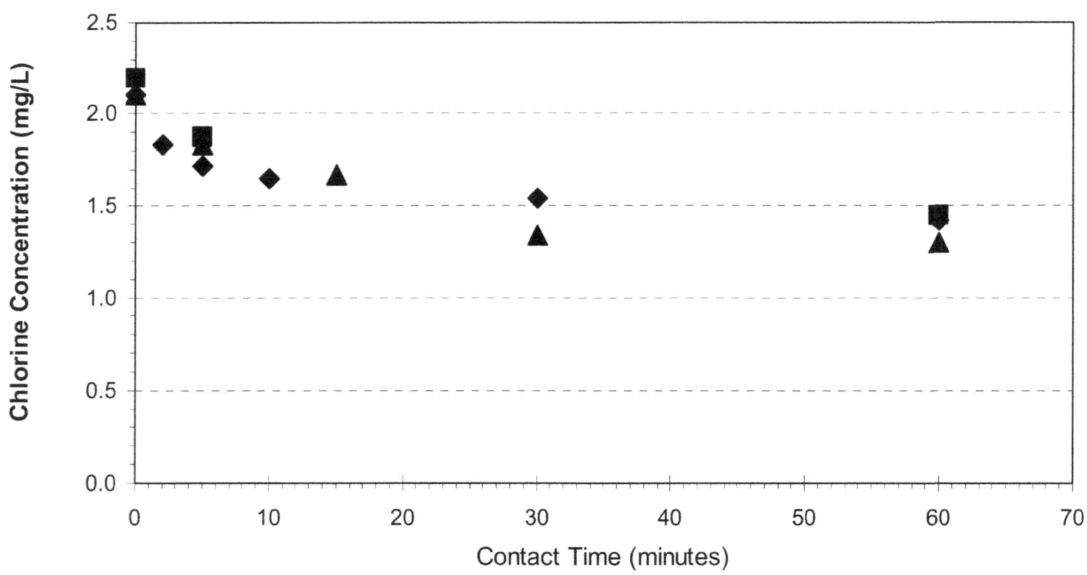

(B)

Figure 6.3 *Streptomycetes* Survival Ratio (A) and chlorine depletion (B) at pH 5 with free chlorine

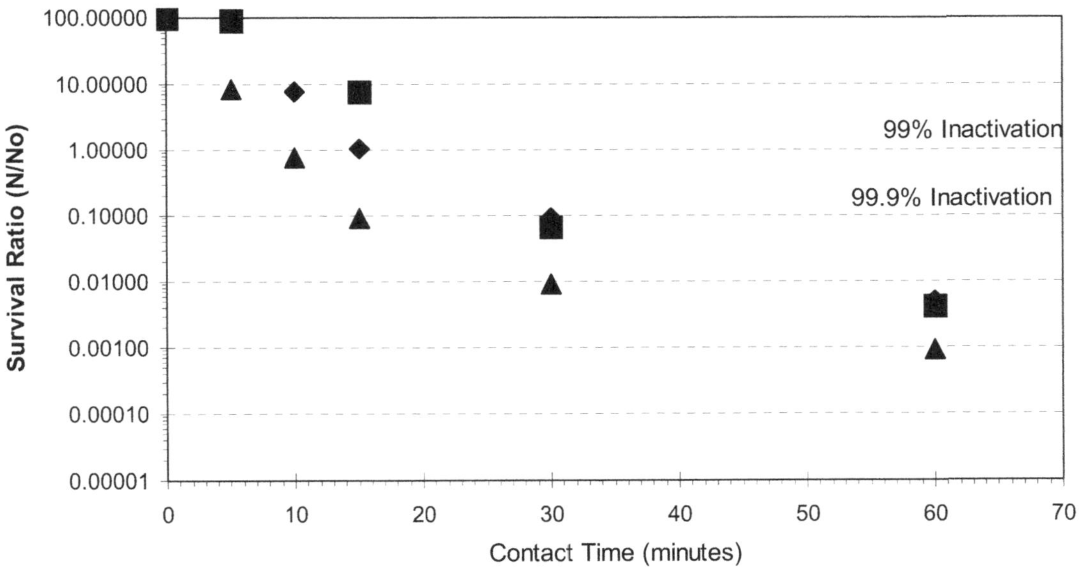

(A)

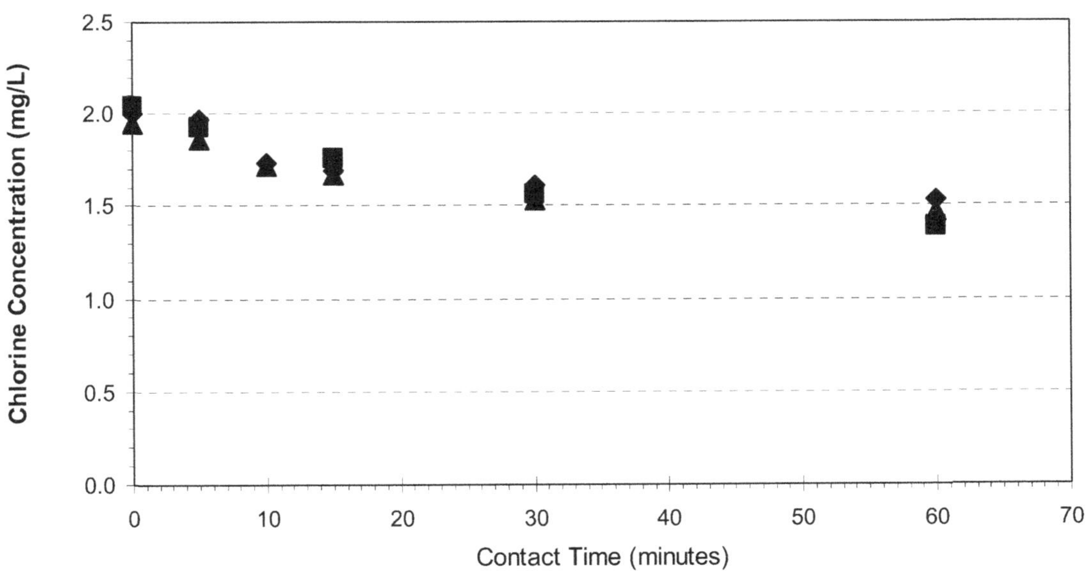

(B)

Figure 6.4 *Streptomycetes* Survival Ratio (A) and chlorine depletion (B) at pH 7 with free chlorine

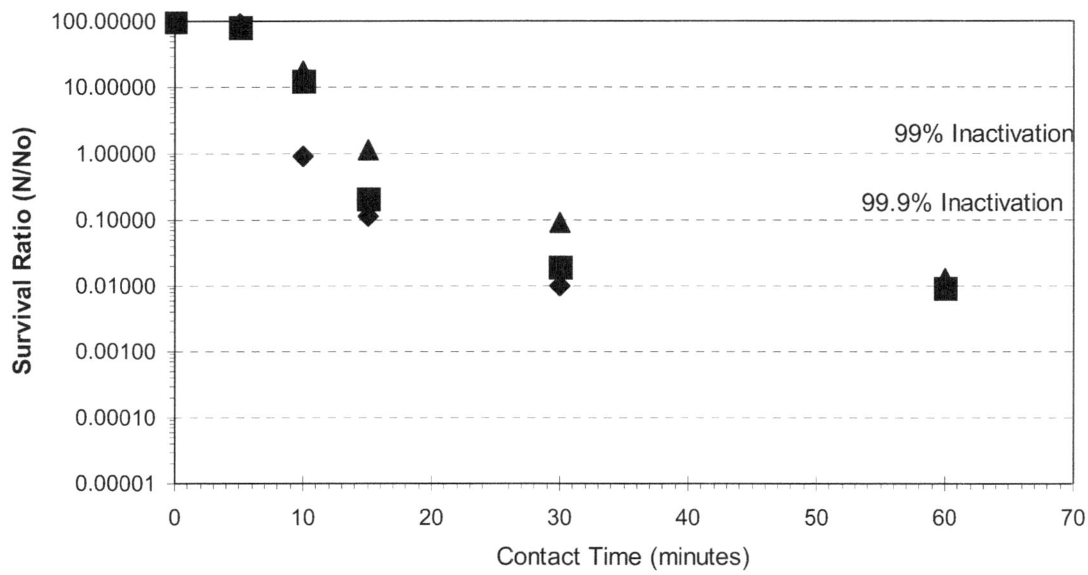

(A)

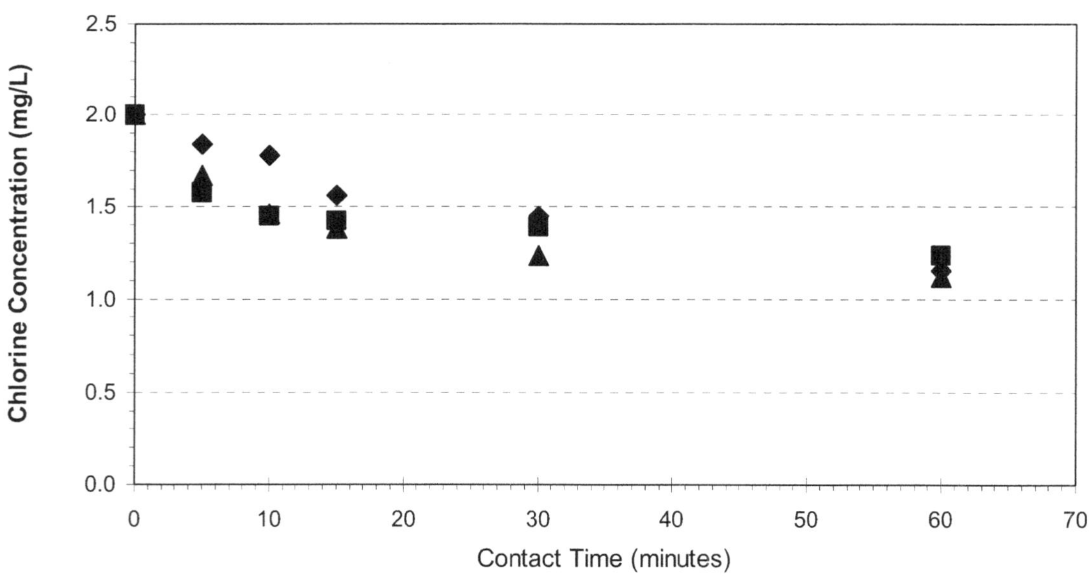

(B)

Figure 6.5 ***Streptomycetes*** **Survival Ratio (A) and chlorine depletion (B) at pH 9 with free chlorine**

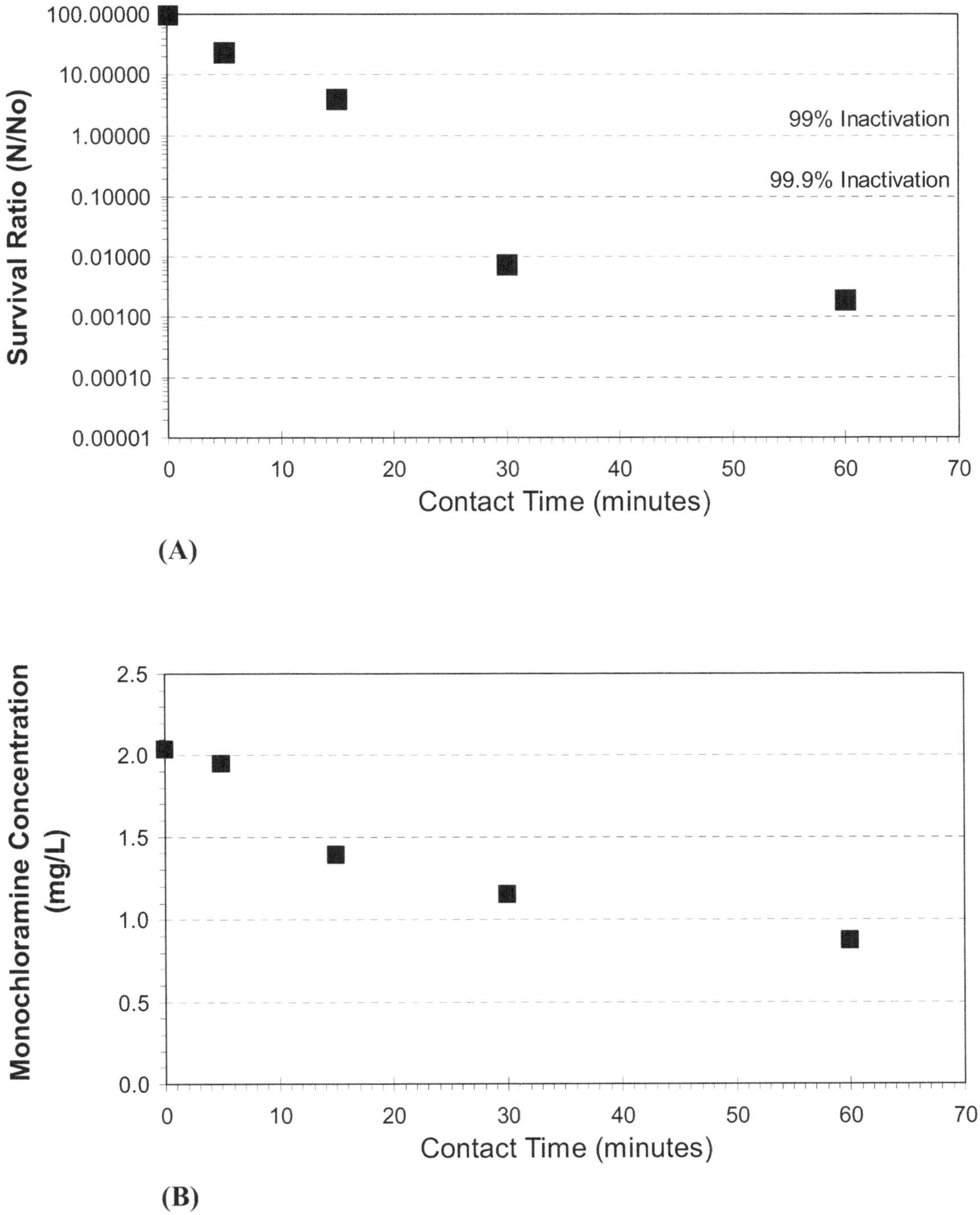

Figure 6.6 *Streptomycetes* Survival Ratio (A) and monochlorine depletion (B) at pH 8.5 with monochloramine

CHAPTER 7
BIOFILMS IN WATER DISTRIBUTION SYSTEMS

CITY OF PHOENIX DISTRIBUTION SYSTEM

One odor-producing organism was isolated from the distribution system samples collected in the baseline monitoring effort conducted on April 16, 2002. The organism was identified as Actinomycete and was originally isolated on R2A agar at 35° C for 72 hours.

The isolated colony was inoculated to several Yeast-Malt-Extract Agar (YMEA) slants for maintenance at 28° C. Subsequent testing was adding 10 mL of sterile water to the slants and gently agitating growth surfaces with an inoculating loop. Two mLs of this water were withdrawn and inoculated into a Fernbach flask containing 500 mLs of Romano-Safferman media with mannitol. Flasks were incubated at 28° C for seven days. After seven days, the inoculated media solutions were filter sterilized by passing the solution through a 0.45 μm filter. This solution was subsequently analyzed using the SPME method. Geosmin was detected at 13 ng/L.

Additional activities were coordinated through the Water Distribution Division for the City of Phoenix. The laboratory received sections of pipe removed from the distribution system as part of a valve replacement job. The pipe was submerged in water and transported to the lab. When received by the lab, staff learned that the pipe section had been stored in this water for over two days before its delivery. Nonetheless, the pipe section was examined and photographed. Initial examination of the pipe showed no visible biofilm growth, pipe corrosion or pipe tuberculation. The interior pipe surface was scraped and the material was examined under a microscope. No biofilm organisms were observed.

Because of the length of time between removal from the distribution system and delivery to the laboratory, staff sought alternate approaches to obtaining samples. Discussions with Water Distribution personnel led to a recommendation that samples be obtained from valve tapping crews. The crews remove four to six inch pipe sections from the distribution system whenever new valves are placed into service.

Two coupons removed from concrete pipe at different locations were collected. The coupons were transported directly to the lab, examined and photographed. The examination again revealed no indication of biofilms. Staff obtained field samples of cast-iron pipe and those results are discussed in a later section.

Rehabilitation of Reservoir number one at the Deer Valley WTP occurred late August 2003. The reservoir was drained to within six inches above the sediment. Samples were taken in Colilert bottles from the top one-half inch of sediment. Five samples were collected, at various locations throughout the reservoir. The samples were plated for isolation of organisms (actinomycetes) which produce T&O compounds. Table 7.1 exhibits the results.

Despite attempts to obtain biofilms from the water distribution systems, no good cultures were obtained. The City of Phoenix has been maintaining a 2.5 $mgCl_2/L$ residual leaving the WTPs since fall 2001. In 2002 the City of Phoenix conducted a study investigating the use of chloramines and the potential for nitrification by biofilms. In that study, old cast-iron pipes were collected and used in the study, and spiked with water from southern California to obtain a culture of nitrifying bacteria.

ANALYSIS OF ACTINOMYCETES IN BIOFILMS FROM PIPES RECOVERED FROM FIELD

Experiments were performed to investigate the actinomycetes concentrations in biofilm growth in the city water distribution system. The actinomycetes concentration in the biofilm samples collected from the interior surface of coupon were analyzed and compare with the actinomycetes population in model distribution system.

The cast-iron coupons were recovered from the field during the month of August 2005 by City of Tempe personnel. The sampling protocol consisted of scraping half of the surface area (241.6cm^2, 249.4cm^2 and 112.26cm^2, respectively) on both sides of the pipe coupon. The biomass from both sides of the coupon were then combined resulting in a total sample area of 120.8cm^2, 124.7cm^2 and 56.13cm^2 for the first, second and third sample, respectively. After scraping was complete, the sampled area was carefully and thoroughly washed with 50ml-autoclaved water to completely remove the biofilm. The biofilm samples were collected in sterile 50ml centrifuge tubes. The tube was vortexed for five minutes to break down the biofilm chunks and to homogenize the samples. The samples were serially diluted (using 10 folder dilution series) and 1ml of each dilution series was plated in duplicate.

The biofilm samples from the three coupons made up of cast-iron were analyzed for actinomycetes and MIB/Geosmin concentrations to evaluate the real world scenarios (Table 7.13). The actinomycetes concentrations in the three biofilms were 6.7 x 10^5, 1.3 x 10^6 and 1.2 x 10^{5}CFU/cm^2, respectively; the MIB concentrations were 2.8, < 2.0 and 50.6 ng/L, respectively; the Geosmin concentrations were < 2.0, 3.6 and 2.6 ng/L, respectively.

The water samples containing these coupons were also analyzed for actinomycetes and MIB/Geosmin concentrations (Table 7.14). The actinomycete concentrations in the three water samples were 1.5 x 10^5 CFU/mL and 4.4 x 10^5 CFU/mL, respectively. The MIB concentrations of 5.8 ng/L and 12.2 ng/L were recorded in the water surrounding the biofilm, respectively. The Geosmin concentrations in these water samples were recorded below detection limit.

The actinomycetes data collected from the stagnation experiment in model distribution system was compared with the data obtained from the city's distribution system. Actinomycetes concentrations were found to be similar in biofilm samples in the city's and in the model distribution systems, averaging at 6.8 x 10^5 and 7.6 x 10^5 CFU/cm^2, respectively. The data provide qualitative indications of effects that could be expected in a full-scale system.

PIPELOOP EXPERIMENTATION

MIB and Geosmin Stability and Depletion Rate

At the initial stage of this study experiments were performed to investigate the decay/dissipation rate of MIB and Geosmin in model drinking water distribution systems. In the preliminary experiments, low concentrations of these compounds were spiked in the distribution systems to mimic a real world scenario. A high and rapid loss in the concentration of geosmin was observed in the model distribution system, which may be due to degradation, sorption, or other causes. Based on these preliminary results it was concluded that geosmin would be of little use to study the stability of T&O compounds in drinking water distribution over an extended period of time. Therefore, MIB was selected for in-depth study of the behavior of T&O compounds in the model distribution system. The focus of extensive investigation was: to investigate the effect of short duration changes in water quality on the MIB depletion rate in the distribution system; to simulate a microbial (actinomycetes) intrusion event in the distribution system; to investigate the effect of source water change on the indigenous actinomycetes population to produce T&O compounds; and to determine any effects of variation in disinfectant-concentrations on the release of T&O compounds by actinomycetes population.

The MIB depletion experiments were performed in duplicate. In the first experiment, cast-iron and PVC pipeloop systems were spiked with 8.2 ng/L and 7.1 ng/L, respectively. The concentrations represent a real world scenario. However these low concentrations were not sufficient to elicit a good depletion curve. For the second experiment, cast-iron and PVC loops were spiked with 107.2 ng/L and 99.5 ng/L, respectively. The MIB and Geosmin concentrations in the tap water coming into our model system were monitored on a weekly basis and data is presented in Tables 7.2 and 7.3. The detectable daily concentrations of MIB measured during May 19 – August 5 were 2.0, 2.2, 2.3, 10.1, 7.0, 5.2, 7.2, 4.8, and 14.4 ng/L.

Depletion of MIB at Low Concentration in the Model Distribution Systems

To investigate the depletion rate of MIB at low concentration, 8.2 ng/L and 7.1 ng/L of MIB were spiked in cast-iron and PVC pipeloop systems, respectively. Figure 7.1 shows the depletion rate at low concentrations of MIB spiked in cast-iron pipeloop. MIB concentration (ng/L) is plotted on y-axis, and sampling time (hours) is plotted on x-axis. A good fit linear regression ($y = -0.1327x + 8.0584$, $R^2 = 0.9839$) of MIB was exhibited during the initial 24 hours. The concentrations of 8.2 ng/L and 4.9 ng/L were observed at hours 0.25 and 24, respectively. The ratio of remaining concentration to the original concentration was 0.61. Depletion rates were determined by assuming a first-order fit of the data using Microsoft Excel.

Figure 7.2 shows the depletion rate of MIB spiked in PVC pipeloop. MIB concentration (ng/L) is plotted on y-axis, and time (hours) is plotted on x-axis. MIB depletion trend in PVC was similar to the one observed in cast-iron. MIB depletion in PVC pipeloops presented a good fit linear regression ($y = -0.0572x + 7.2094$, $R^2 = 0.9754$) during the initial 24 hours. The MIB concentrations of 7.1 ng/L and 5.8 ng/L were observed at hours 0.25 and 24, respectively. The ratio of remaining concentration to the original concentration was 0.80. Therefore, it can be concluded that during the initial 24 hours the depletion rate in cast-iron was faster than in PVC.

A similar trend was observed after the initial 24 hours. In cast-iron, it took one and a half days for MIB to reach its half-life estimated from the graph and seven days to decline below the detection limit (2.0 ng/L), while four days and seventeen days were needed for PVC (Figure 7.3).

To investigate the depletion rate of Geosmin, 4.9 ng/L and 5.2 ng/L of Geosmin were spiked in cast-iron and PVC pipeloop systems, respectively. Figure 7.4 shows the depletion rate of Geosmin spiked in cast-iron (4.9 ng/L) and PVC (5.2 ng/L) pipeloops. Geosmin concentration (ng/L) is plotted on the y-axis, and sampling time (hour) is plotted on x-axis. The depletion of Geosmin fits the trend line $y = 4.2305x^{-0.2234}$, $R^2 = 0.9431$ for cast-iron and $y = 4.662x^{-0.1808}$, $R^2 = 0.9427$ for PVC during the initial 24 hours. In cast-iron, it took 24 hours for Geosmin to decline below the detection limit (2.0 ng/L), while 72 hours were needed for the PVC pipeloop. Based on the data, the depletion rate of Geosmin in cast-iron was faster than PVC during the initial 24 hours. Due to this rapid depletion rate Geosmin was not utilized in subsequent pipeloop experiments.

Depletion of MIB and Geosmin at High Concentration in the Model Distribution Systems

To investigate the depletion rate of MIB at high concentration, the cast-iron and PVC pipeloop systems were spiked with 107.2 and 99.5 ng/L, respectively. Water samples for MIB were collected for 80 days (Figure 7.5). MIB concentration (ng/L) is plotted on y-axis, and sampling time (day) is plotted on x-axis. In cast-iron, 107.2, 58.7, 61.3, 54.0, 61.6, 54.7 ng/L of MIB concentrations were observed at day zero, one, two, three, four, and five, respectively. In PVC, MIB concentrations of 99.5, 67.2, 66.1, 61.2, 53.6, 65.1 ng/L were observed at day 0, 1, 2, 3, 4, and 5, respectively. After ten days, MIB concentration declined fairly slowly in both systems. MIB depletion rates ranged from 48.5 ng/L/d to 0.03 ng/L/d in cast-iron, and 32.3 ng/L/d to 0.18 ng/L/d in PVC over a 80-day period. In cast-iron, it took three days for MIB to reach its half-life estimated from the graph, while four days were needed in PVC. MIB concentrations were higher than Geosmin concentrations for > 95% of the samples. Based on the trend exhibited by MIB depletion period, the MIB depletion in distribution systems can be divided into three discrete stages: the initial 24 hours, day 1 to day 5, and day 6 to day 80. The depletion in the first stage is exhibited in Figure 7.6. The ratio of remaining concentration to the original concentration (C/Co) is plotted on the y-axis; the sampling time (day) is plotted on the x-axis. Two linear regression ($y = -0.0112x + 0.9708$, $R^2 = 0.8921$ for PVC and $y = -0.0142x + 0.9591$, $R^2 = 0.7041$ for cast-iron) of MIB were exhibited during the initial 24 hours. Based on the data shown above, in cast-iron, MIB sharply decreased during the initial 24 hours; it reached almost half of the initial concentration. However, the concentrations increased during the 12 to 18 hour period. The fastest depletion rate occurred over the period of the last 6 hours (18 - 24hours). In PVC, the concentration declined to two-thirds of the initial concentration.

The changes in MIB concentrations from day 1 to 5 (the second stage) are exhibited in Figure 7.7. The ratio of remaining MIB to original concentration (C/Co) is plotted on y-axis; the sampling time in unit of day is plotted on x-axis. Two linear regression ($y = -0.0168x + 0.6799$, $R^2 = 0.2275$ for PVC and $y = -0.0072x + 05632$, $R^2 = 0.7041$ for cast-iron) of MIB were exhibited during the initial 24 hours. Based on the data shown above, neither sharp increase nor decrease in the concentration was recorded. Comparing MIB concentration at day one to that at day five for both pipeloops, the values were almost at the same level, which were 0.55 and 0.51 for cast-iron, 0.68 and 0.65 for PVC, respectively. Thus it could be concluded that after experiencing rapid depletion over the initial 24 hours, the MIB went through a relatively stable period during the subsequent four days.

The changes in MIB concentrations from day 6 to day 80 (the third stage) are exhibited in Figure 7.8. The ratio of remaining MIB to original concentration was plotted on y-axis; the sampling time (day) was plotted on x-axis. Based on the data shown above, the depletion of MIB fits the trend line $y = 2.4577x^{-0.6833}$, $R^2 = 0.891$ for cast-iron, $y = 1.9785x^{-0.4012}$, $R^2 = 08486$ for PVC. However, it was observed that MIB concentrations increased at certain days; this phenomenon showed sporadic MIB production. For instance, the increases of 5.4 ng/L, 2.7 ng/L for cast-iron occurred from days 18 - 19, days 38 - 39, respectively; the increases of 3.1 ng/L, 3.7 ng/L, 4.7 ng/L, 8.4 ng/L, 3.5 ng/L for PVC occurred from days 7 to 8, days 27 to 28, days 29 to 30, days 38 to 39, and days 44 to 45, respectively. The changes in MIB concentrations in pipe loops may be due to MIB degradation and fresh MIB production as a result of some chemical, physical and biological reactions within the reservoir. It is important to point out that no water was entering the pipeloop system during those days.

EFFECT OF WATER CIRCULATION AFTER STAGNATION

Effect of Restarting Water Circulation after Stagnation

Each system was seeded with 202 mL actinomycetes spiking-solution with concentration of 5.7 x 10^7 CFU/mL. After the initial set of experiments the water circulation in the distribution system was stopped for 12 days, and during this time period the whole system acted as a dead end. After 12 days stagnation, water circulation in both systems as restarted and water samples were collected from dead-end, flowing water and reservoir sampling points, 5 minutes after re-starting the water circulation. Thereafter, biofilm samples were collected. Actinomycetes and MIB concentrations in various parts of the distribution systems are presented in Table 7.4 and 7.5. The maximum measured concentration of geosmin produced by actinomycetes was 3.6 mg/L.

Effect of Stagnation on the Survival of Actinomycetes in Flowing Water

Cast-iron and PVC systems were inoculated with actinomycetes stock, and then water circulated for 4 days. Thereafter, the system was shut down for 12 days. Water samples were analyzed for actinomycetes and data is presented in Figures 7.9 and 7.10.

Actinomycetes concentration (CFU/mL) is plotted on the y-axis and time (day) is plotted on the x-axis. The total number of actinomycetes added in each system was approximately 1.3 x 10^{10}. Four days later, actinomycetes in flowing water decreased in PVC and cast-iron; they were attached on the surface of biofilm and forming the majority of microorganisms in biofilms. After 12 days stagnation, actinomycetes concentration was increased immediately after re-starting the water circulation. The increases were 2 logs for cast-iron, and 1 log for PVC. It appears that disturbances in the pipeline that resulted from 5 minutes' flushing may then release deposits containing actinomycetes into flowing water.

Effect of Stagnation on MIB Concentration in Flowing Water

Cast-iron system was inoculated with actinomycetes stock, and then water circulated for 4 days to stabilize its population. Thereafter, the system was shut down for 12 days. Water samples were analyzed for MIB and data is presented in Figure 7.11.

At day 12 the MIB sample was collected 5 minutes after restarting water circulation. In cast-iron, MIB concentrations of 4.3 ng/L and 33.5 ng/L were observed in flowing water at day 0 and 12, respectively; 5.1 ng/L and 36.5 ng/L were observed in dead-end at day 0 and 12, respectively; 4.6 ng/L and 43.2 ng/L were observed in reservoir at day 0 and day 12, respectively. Briefly, 0.89 log, 0.85 log and 0.97 log increase in actinomycetes were observed in flowing water, dead-end and reservoir, respectively. During prolonged stagnation period actinomycetes tend to settle down in biofilms. Significantly higher concentrations (25 ng vs. 1000 ng per liter) of MIB are produced by the microorganisms present in sediment compared to the microorganisms in flowing water (Izaguirre et al., 1982). It can be concluded that restarting water circulation after 12 days stagnation may result in the release of high concentration of MIB in dead-end, flowing water and reservoir.

The PVC system was inoculated with actinomycetes stock, and then water circulated for 4 days. Thereafter, the system was shut down for 12 days. Water samples were analyzed for

MIB and data is presented in Figure 7.12. At day 12 the MIB sample was collected 5 minutes after restarting water circulation. In PVC, MIB concentrations of 16 ng/L and 44.5 ng/L were observed in flowing water at day 0 and 12, respectively; 17.3 ng/L and 43.9 ng/L were observed in dead-end at day 0 and 12, respectively; 16.8 ng/L and 53.5 ng/L were observed in reservoir at day 0 and day 12, respectively. After 12 days stagnation, 0.44 log, 0.40 log and 0.50 log increase in MIB were observed in flowing water, dead-end and reservoir, respectively. It can be concluded that restarting water recirculation after 12 days stagnation results in releasing of high concentrations of MIB in dead-end, flowing water and reservoir.

Comparing cast-iron with PVC for MIB fluctuation in various sections of distribution system, the two systems showed similar increasing trend in MIB concentration in flowing water. But it appeared that cast-iron provides more favorable biofilm conditions for production and release of MIB in the flowing water than PVC. Direct correlation was observed between actinomycetes and MIB concentrations.

The effect of stagnation on MIB concentration in flowing water is shown in Figure 7.13. Prior to seeding actinomycetes, MIB concentrations were 7.2 ng/L and 20.8 ng/L in cast-iron and PVC, respectively. During 12 days stagnation, substantial increase in MIB was observed for both PVC and cast-iron with no obvious differences between flowing water and dead-end. The increase in MIB concentration was correlated to the increase in actinomycetes population in distribution systems.

The average increase rate of MIB concentration in flowing water was 2.4 ng/L/day in both pipeloops. However, in the dead-end of cast-iron pipeloop, MIB concentration increased at a rate of 2.7 ng/L/day, which was a slightly faster than that in PVC with 2.2/ng/L/day. It appears that cast-iron provides more favorable conditions for the production of MIB than PVC.

Effect of Stagnation on Actinomycetes Population in Biofilms

The effect of stagnation on actinomycetes population in biofilm is exhibited in Figure 7.14. Actinomycetes concentration in distribution systems is plotted on the y-axis and sampling time is plotted on the x-axis. Biofilm samples were collected one day before stopping water circulation. During the 12 days stagnation, the system's biofilm switched from aerobic to anaerobic condition. After 12 days stagnation, the systems were restarted for 5 minutes and the biofilm samples were collected again.

Actinomycetes concentration was 1-log increased, from $14/mm^2$ to $176/mm^2$ of biofilm in PVC, while a 1.2-log decrease, from $7625/mm^2$ to $390/mm^2$ of biofilm in cast-iron was observed. In cast-iron, stagnation resulted in superficial deposition of actinomycetes on biofilm. Actinomycetes present on the biofilm can easily become suspended in flowing water during flushing. In PVC, actinomycetes were embedded in biofilm and could not easily be resuspended.

Comparison of Stagnation Effect on Actinomycetes in Flowing Water and Biofilms

The effect of stagnation on actinomycetes in flowing water and biofilm is exhibited in Figure 7.15. Actinomycetes concentration is plotted on the y-axis. In cast-iron, 12 days stagnation resulted in 1.07 log and 1.46 log increases in flowing water and dead-end, respectively, while 1.29 log decrease was observed in $biofilm/mm^2$. In PVC, 12 days stagnation resulted in 0.36 log, 0.28 log, and 1.10 log increase in actinomycetes in flowing water, dead-end and $biofilm/mm^2$, respectively.

Effect of Source Water Replacement from Surface to Ground Water

Ground water was collected from Scottsdale Well Site # 4E on March 17, 2005. Prior to pumping the 35 liter groundwater to each reservoir, electrical pump in cast-iron pipeloop system was cleaned to ensure similar flow rate & pressure conditions were in both systems. The temperature in both systems was maintained at 24° C. Ground water samples were analyzed for actinomycetes concentration and MIB levels to determine the background concentrations. Actinomycetes stock was seeded into reservoir, and water samples were collected from flowing water, dead-end and the reservoir sampling points. The sampling schedule was strictly followed over 15 days. On day 7, 15 liters of groundwater was added into each pipeloop to reach the maximum capacity. Despite the additional dilution factor, no significant change in actinomycetes and MIB was observed.

Effect of Ground Water on the Survival of Actinomycetes in the Flowing Water

In PVC, the fluctuations of actinomycetes concentrations in flowing water over time were persistent during the entire time period (Figure 7.16). This phenomenon indicates that the pilot PVC pipeloop distribution system has little effect on either production or loss of actinomycetes populations. In PVC supplied with groundwater, actinomycetes concentrations of 6.72, 6.51, 6.61, 6.43, 6.79, 6.54, 6.45, and 5.81 log were recorded at day 0, 2, 4, 6, 8, 10, 12, and 14, respectively. Similar trends were observed for dead-end and reservoir sampling points. At day 15, the final day, the log concentrations of actinomycetes were 5.86, 5.52, and 5.73 in flowing water, dead-end and the reservoir, respectively. They all were lower than the concentrations at day zero for each corresponding sampling port.

In cast-iron, the fluctuations in actinomycetes concentrations in flowing water were observed during the initial 7 days (Figure 7.17). Actinomycetes concentrations were 5.68, 5.00, 5.19, 4.82, and 5.34 log at day 0, 2, 3, 5 and 7, respectively; while slight increases in actinomycetes population were observed over the last eight days. The increase in actinomycetes concentrations were 4.47 and 4.66 log at day 10 and 15, respectively. Similar trends were observed in dead-ends and reservoirs.

Effect of Ground Water on MIB Production in the Flowing Water

In PVC, MIB concentrations were increased over the course of experiment (Figure 7.18). This phenomenon indicates that the overall net production rate of MIB was higher than the overall net loss rate in PVC. The MIB levels in the flowing water were 5.5, 5.1, 6.8, 9.5, 11.7, 11.0, 13.4, and 17.2 ng/L at day 0, 2, 4, 6, 8, 10, 12, and 15, respectively. The fluctuations in MIB concentrations always occurred in each sampling port but were not similar to each other. However, the levels of MIB reached 17.5 ng/L on day 14 and then it started to decrease. It is interesting to note that during the initial five days, flowing water and dead-end exhibited almost the same level of MIB at each day; and also, the MIB levels in reservoir were higher than those in flowing water and dead-end samples over the course of the experiment. However, the concentrations of MIB in all the sampling ports reached the same level after 15 days.

In PVC, MIB levels peaked at day fourteen, which had 0.33 log, 0.5 log, 0.43 log increase for reservoir, flowing water and dead-end, respectively (Figure 7.19).

Figure 7.20 presents MIB concentration in groundwater in cast-iron pipeloop. In cast-iron, after 15 days, no obvious overall increase or decrease in MIB concentration was noted for the three selected sampling ports, and the concentrations were approximately 15 ng/L. For flowing water, the levels of MIB were 11.8, 15.1, 17.6, 16.7, 14.3, 14.5, 16.9, 14.2 ng/L at day 0, 2 4, 6, 8, 10, 12, and 15, respectively. However, on certain individual days, the obvious increase and decrease in MIB were observed; for an instance, from day 6 to 7, MIB levels increased from 8.4 ng/L to 16.8 ng/L for dead-end samples. Similar to PVC pipeloops, the levels of MIB in flowing water and reservoir exhibited small changes during the initial five days. Compared to dead-end water, flowing water exhibited more stable MIB levels over the course of experiment, though MIB fluctuations always occurred.

The results shown in Figure 7.21 represent the changes in MIB concentration that occurred in groundwater in cast-iron pipeloop. It is interesting to note that no matter how the MIB levels fluctuated on a day to day basis, the final levels measured at day 15 returned to the original spiking level since the ratio of remaining MIB to the original MIB concentration approximately equaled zero for each sampling port.

Comparison of Actinomycetes Population in Surface and Ground Water

Figure 7.22 presents the results of actinomycetes survival rate performed during the summer of 2004, to investigate the changes in actinomycetes concentrations in surface water over time.

The concentration of actinomycetes, decreased less rapidly in the reservoir and flowing water compared to the dead-end. Two logs decrease in actinomycetes concentration (from 4.1 x 10^4 to 3.9 x 10^2 CFU/mL) in flowing water over a 7-day period was observed. In the dead-end, actinomycetes concentration increased from 2.4 x 10^4 to 4.6 x 10^4 CFU/mL during the first 24 hours; thereafter a steady decrease in actinomycetes concentration was observed. Overall, two log decrease in actinomycetes concentration in distribution system was observed over a 7-day period. A slight decrease in the concentration of actinomycetes in flowing water occurred with the decrease in flow rate, while in dead-end no such impact was observed.

Apparently, changes in actinomycetes concentrations in the selected sampling ports in PVC supplied by groundwater were different from those using surface water. Actinomycetes concentrations of the groundwater fluctuated with slight decrease over time; while for the surface water, rapid decrease in actinomycetes concentration was observed.

Comparison of MIB in Surface and Ground Water

Figure 7.23 presents the results of MIB survival experiment performed during summer of 2004 to investigate the changes in MIB concentrations in surface water over time.

A rapid decrease in MIB concentrations in PVC pipeloop was observed over time (Figure 7.23). The depletion rates ranged from 0.9 ng/L/d to 9.7 ng/L/d in sampling ports for most sampling days except for two occasions: one was 0.3 ng/L/d increase in flowing water during day 2 and 3, the other was 0.2 ng/L/d increase in dead-end during the first day.

In PVC pipeloop supplied with the ground water, MIB showed a different trend from that in surface water. The MIB concentrations of groundwater increased over time while rapid decrease was observed in surface water.

Effect of Chlorination and Rechlorination

For this experiment, the chlorination was performed using bleach. To collect flowing water samples, a sampling port in a straight horizontal segment of pipe was selected since it represents a main section of water distribution systems. A day prior to the chlorination experiment, each reservoir in cast-iron and PVC pipeloop was seeded with actinomycetes containing $6x10^4$ CFU/mL. Water samples from both systems were collected 1 hour after spiking and immediately analyzed. The initial actinomycetes concentrations were 2.2×10^3 CFU/mL and 2.7×10^3 CFU/mL in cast-iron and PVC, respectively.

The chlorination and rechlorination experiment consisted of three chlorination regimes. Initially, mass dose of free chlorine was maintained at 1.0 mg/L for 24 hours. Then the chlorine injection pump was shut down for 48 hours, which was followed by a chlorination episode in which chlorine was maintained at 0.5 mg/L for 48 hours. The pump was stopped for another 48 hours and in the third chlorination episode, chlorine was maintained at 2.0 mg/L for 24 hours (Figure 7.24).

Background Information on Free Chlorine Levels in Tap Water

Table 7.6 shows the total chlorine and free chlorine concentrations in surface water during the summer of 2004. The free chlorine exhibited in the tap water treated by the city of Tempe ranged from 0.03 mg/L to 0.70 mg/L from June 21 to July 14. The sharp change in chlorine concentration occurred on June 29 and June 30, during which the free chlorine levels increased from 0.10 mg/L to 0.70 mg/L. However, the free chlorine levels in the aeration tank ranged from 0.01 mg/L to 0.04 mg/L throughout the sampling period. It appears that the open tank provides a mechanism to release chlorine from water prior to entering into the model distribution systems.

Maintenance of Chlorine Residual in the Distribution Systems

Accurate maintenance of chlorine residual at a prescribed level in the cast-iron and PVC pipeloops was a difficult task. Although, the chlorine level was monitored and adjusted on a 15-minute basis, fluctuations were difficult to control. In cast-iron system, chlorination seemed to result in increased turbidity, which might have been due to the effect of chlorine on minerals in the water (such as iron) and biofilms. The chlorine consumption rates in cast-iron system were faster than those in PVC (Figures 7.25 – 7.28).

Concentrations of Actinomycetes and MIB in Flowing Water in Cast-iron During Chlorination and Rechlorination

The concentrations of actinomycetes and MIB in flowing water in cast-iron over time are exhibited in Figures 7.29 and 7.30. In Figure 7.29, the actinomycetes concentration is plotted on the left y-axis in CFU/mL and chlorine residual is plotted on the right y-axis in mg/L. Based on the data, it can be concluded that actinomycetes concentrations in flowing water rapidly decreased over first event of chlorination. However, that decrease in actinomycetes population was transient, as an increase in actinomycetes number was observed with the depletion of chlorine level in the system. Rechlorination of the system (at 0.5 mg/L) resulted in rapid decrease in actinomycetes population and release of high-levels of MIB. Over the last 12 hours of the second chlorination event, a rapid decrease in MIB level was recorded, which might be due to the oxidation of MIB. Thereafter, the MIB level presented a slight decrease followed by a slight increase during chlorine depletion period. The third chlorination event at 2.0 mg/L resulted in relatively stable and consistent results. Apparently, this higher disinfection had little effect on the MIB level present in the system. The actinomycetes population in the system exhibited the similar trend over the last two chlorination events. However, the effects of chlorination and rechlorination cannot be considered to be totally independent of each other due to the potential residual impact on the system.

The First Event of Chlorination in Cast-Iron

The concentrations of actinomycetes and MIB in the flowing water in cast-iron over time during first chlorination event and the subsequent depletion of chlorination are presented in Table 7.7 In cast-iron pipeloop with the baseline actinomycetes concentration of 2.2 x 10^3 (CFU/mL), a rapid decrease in actinomycetes concentration was noted right after chlorination started. During the first chlorination event, actinomycetes concentrations of 57, 58, 26, 54, 20, 28, 105, and 174 CFU/mL were recorded at 0.5, 1, 2, 3, 6, 12, 18, and 24 hours, respectively. However, actinomycetes concentrations increased with the depletion of chlorine; the concentrations were 1300, 2380 CFU/mL at 48 and 72 hours, respectively. Under similar conditions MIB levels were 2.3, 3.6, 2.0, 2.0, 2.6, 2.6, 2.3, 2.1 ng/L at 0.5, 1, 2, 3, 6, 12, 18, 24 hours, respectively. However, the MIB levels did not change over the following period during which there was a depletion of chlorine, and they were 2.5 and 2.1 ng/L at 48 and 72 hours, respectively.

For Figure 7.31, actinomcyete concentration is represented on the left y-axis, while MIB concentration and chlorine levels (mg/L) are represented on the right y-axis. In Cast-iron, 1.16-log increase in actinomycetes population was obtained over the period showing chlorine depletion, while little change in MIB level was observed during either the chlorination period or following period of chlorine depletion.

The Second Event of Chlorination in Cast-Iron

For cast-iron, the concentrations of actinomycetes and MIB in flowing water over time during the second chlorination event and the subsequent chlorine depletion period are presented in Table 7.8. In cast-iron pipeloop, the actinomycetes concentration rapidly decreased from 2,380 to 144 CFU/mL right after re-chlorination started. During the second chlorination event, actinomycetes concentrations of 144, 58, 163, 36, 62, 127, 192, 223, 51, 233 CFU/mL were recorded at 0.5, 1, 2, 3, 6, 12, 18, 24, 36 and 48 hours, respectively. With the depletion of chlorine residual, actinomycetes concentrations were 2,380, 950 CFU/mL at 72 and 96 hours, respectively. This phenomenon indicates that short periods of chlorine depletion could have an effect on actinomycetes production. Under similar condition of re-chlorination (0.5 mg/L), MIB levels of 2.1, 2.4, 2.4, 2.5, 22.7, 21.1, 23.8, 23.3, 24.1, 13.4 ng/L were recorded at 0.5, 1, 2, 3, 6, 12, 18, 24, 36 and 48 hours, respectively. The substantial increases in MIB levels from 2.5 ng/L to 22.7 ng/L were observed during hours 75-78. However, the MIB levels didn't exhibit substantial increase or decrease over the chlorine depletion period, and they were 10.1 and 14.2 ng/L at 72 and 96 hours, respectively.

Figure 7.32 shows the results of actinomycetes and MIB concentrations in cast-iron during the second chlorination event. Actinomycetes concentration is represented on the left y-axis, while MIB concentration and chlorine level (mg/L) are represented on the right y-axis. In cast-iron, an overall 0.6 log increase in actinomycetes populations was recorded over the chlorine depletion period, while nearly 1.0 log increase in MIB level was observed during hours 75 - 78, which indicated that the rechlorination (0.5 mg/L) resulted in release of high levels of MIB.

The Third Event of Chlorination in Cast-Iron

Concentrations of actinomycetes and MIB in flowing water in cast-iron during third chlorination event are presented in Table 7.9.

In cast-iron pipeloop, the actinomycetes concentration rapidly decreased from 950 CFU/mL to 50 CFU/mL right after re-chlorination (at 2.0 mg/L) started. Actinomycetes concentrations of 50, 30, 9, 35, 24, 27, 17 CFU/mL were recorded at 0.5, 1, 2, 3, 6, 12, and 24 hours, respectively. During the period chlorine was being depleted, actinomycetes concentration of 528, 4600, 1250 CFU/mL were observed at 48, 72 and 96 hours, respectively. Under similar condition, MIB concentrations of 16.3, 18.5, 19.6, 14.0, 15.7, 16.3 1 ng/L were recorded at 1, 2, 3, 6, 12, 24 hours, respectively. The MIB levels didn't exhibit substantial increase or decrease during chlorine depletion period and they were 16.2, 15.5 and 14.1 ng/L at 48, 72 and 96 hours, respectively.

Figure 7.33 shows actinomycetes and MIB concentrations in cast-iron during the third event of chlorination. Actinomycetes concentration (CFU/mL) is represented on the left y-axis, while MIB concentration and chlorine level (mg/L) are represented on the right y-axis. A 1.9-log increase in actinomycetes population occurred during chlorine depletion period, while little change in MIB level was observed during either the chlorination or chlorine depletion period.

Concentrations of Actinomycetes and MIB in Flowing Water in PVC During Chlorination and Rechlorination

The concentrations of actinomycetes and MIB in flowing water in PVC are exhibited in Figures 7.34 and 7.35. Actinomycetes concentration in water rapidly decreased over first event of chlorination (Figure 7.35). However, the actinomycetes concentration increased with the depletion of chlorine level in the system. Rechlorination of the system (at 0.5 mg/L) again resulted in rapid decrease in actinomycetes population. Actinomycetes population again increased during the second chlorine depletion period (120 - 168 hours) (Figure 7.35). Repeated re-chlorination of the system at higher residual concentration had no significant difference on the actinomycetes population. The MIB levels in PVC and cast-iron exhibited the similar trend over the entire experiment period. However, the effects of chlorination and rechlorination cannot be considered independent of each other due to the potential residual impact on the system.

The First Event of Chlorination in PVC

Concentrations of actinomycetes and MIB over time during the first chlorination event are shown in Table 7.10. In cast-iron pipeloop, the actinomycetes concentration rapidly decreased from 2,720 to 227 CFU/mL within 30 minutes of chlorination (Table 7.10). During the first chlorination event, concentrations of 227, 43, 72, 44, 3, 5, 10 and 15 CFU/mL were recorded at 0.5, 1, 2, 3, 6, 12, 18, and 24 hours, respectively. However, the actinomycetes concentrations again increased during the chlorine depletion period and were 1440, 2540 CFU/mL at 48 and 72 hours, respectively. Under similar conditions MIB levels were 2.6, 2.0, 2.0 3.6, 2.0, 2.0, 2.0, and 2.0 ng/L at 0.5, 1, 2, 3, 6, 12, 18, 24 hours, respectively. However, MIB levels didn't exhibit change during the subsequent period of chlorine depletion, and they were 2.3 and 2.0 ng/L at 48 and 72 hours, respectively. Actinomycetes concentration is represented on the left y-axis, while MIB concentration and chlorine level (mg/L) are represented on the right y-axis (Figure 7.36). In PVC, 2.2-log increase in actinomycetes population was observed during chlorine depletion period, while little change in MIB level was observed during either the chlorination or chlorine depletion period.

The Second Event of Chlorination in PVC

Concentrations of actinomycetes and MIB over time during second event chlorination are presented in Table 7.11. In PVC pipeloop, the actinomycetes concentration rapidly decreased from 2,540 to 600 CFU/mL within 30 minutes (Table 7.11). During the re-chlorination event (0.5mg/L) actinomycetes concentrations of 600, 927, 318, 80, 280, 33, 41, 80, 12, and 111 CFU/mL were recorded at 0.5, 1, 2, 3, 6, 12, 18, 24, 36 and 48 hours, respectively. With the depletion of chlorine residual, actinomycetes concentration of 1620, 4000 CFU/mL were observed at 72 and 96 hours, respectively. This phenomenon indicates that actinomycetes regrow during chlorine depletion period. Under similar conditions MIB levels of 2.0, 2.0, 2.3, 20.5, 22.4, 21.7, 24.6, 21.7, 21.3, and 17.9 ng/L were recorded at 0.5, 1, 2, 3, 6, 12, 18, 24, 36, and 48 hours, respectively. A substantial increase in MIB levels was observed, which changed from 2.3 ng/L to 20.5 ng/L during hours 74-75. However, the MIB levels didn't change over the chlorine depletion period, and they were 14.9 and 13.2 ng/L at 72 and 96 hours, respectively.

Figure 7.37 shows the results of actinomycetes and MIB concentrations in PVC during the second chlorination event. Actinomycetes concentration (CFU/mL) is represented on the left y-axis, while MIB concentration and chlorine levels (mg/L) are represented on the right y-axis. In PVC, 1.5 log increase in actinomycetes population was recorded during the chlorine depletion period, while 0.95-log increase in MIB concentration was observed during hours 74 - 75. It could be concluded that the re-chlorination (0.5 mg/L) resulted in release of high level of MIB.

The Third Event of Chlorination in PVC

Concentrations of actinomycetes and MIB in flowing water in PVC pipeloop during the third chlorination event are presented in Table 7.12. In PVC pipeloop, actinomycetes concentrations rapidly decreased from 4,000 to 158 CFU/mL within 30 minutes of re-chlorination at 2.0 mg/L. Actinomycetes concentrations of 158, 16, 30, 65, 21, 33, and 39 CFU/mL were recorded at 0.5, 1, 2, 3, 6, 12, and 24 hours, respectively. Subsequently, during the chlorine depletion period, actinomycetes concentrations of 640, 1008, 3640 CFU/mL were recorded at 48, 72 and 96 hours, respectively. Under similar conditions MIB concentrations of 17.7, 19.4, 17.6, 19.2, 16.7, 14.8 ng/L were observed at 1, 2, 3, 6, 12, 24 hours, respectively. The MIB levels didn't change substantially during chlorine depletion period, and they were 15.4, 13.2, and 15.2 ng/L at 48, 72 and 96 hours, respectively.

Figure 7.38 shows actinomycetes and MIB concentrations in PVC pipeloop during the third chlorination event. Actinomycetes concentration is represented on the left y-axis, while MIB concentration and chlorine levels (mg/L) are represented on the right y-axis.

In PVC pipeloop, 1.97-log increase in actinomycetes concentration was obtained during the chlorine depletion period, whereas MIB concentration didn't change during the chlorination or chlorine depletion period.

SUMMARY

The concentration of actinomycetes and T&O compounds in the distribution system are affected by factors such as pipe material, chlorination, and stagnation. The depletion rate of MIB was greater in cast iron pipeloop compared to the PVC pipeloop. The fastest depletion in both systems was observed during the initial 24 hours after their introduction in the system. Stagnation resulted in increase in the concentration of MIB in water and the stagnation effect was more pronounced in cast iron system that in the PVC system. The concentrations of actinomycetes in both pipe loops were inversely correlated with the chlorine residual in the systems. Specifically, concentrations of actinomycetes and T&O compounds are affected by:

- Pipe material
- Chlorination
- Stagnation
- After experiencing rapid depletion over initial 24 hours, MIB goes through relatively stable period.
- Disturbances in pipeloop that result in flushing may then release deposits containing actinomycetes into flowing water.

Table 7.1
Results from reservoir sample

Location	Actinomycetes (CFU/mL)	HPC-28C 7 days (CFU/mL)
Reservoir midpoint	< 1	140
Mouth of South Channel	< 1	2300
Inside Channel	< 1	5300
Mouth of Channel	< 1	40000
Splitter Box	< 1	140000

Table 7.2
Weekly-monitoring of MIB/Geosmin concentration (ng/L) in the city of Tempe tap water and the pipeloops selected sampling ports during the summer of 2004

Sampling Time	05/19/04		05/27/04		06/03/04	
Sampling Ports	**MIB**	**Geosmin**	**MIB**	**Geosmin**	**MIB**	**Geosmin**
Tap Water	<2.0	<2.0	2.2	2.0	2.3	<2.0
After Aeration Tank	<2.0	<2.0	<2.0	<2.0	2.0	<2.0
Flowing Water (PVC)	2.4	<2.0	N/A	N/A	N/A	N/A
Flowing Water (cast-iron)	2.6	<2.0	N/A	N/A	N/A	N/A
Sampling Time	07/01/04		07/08/04		07/15/04	
Sampling Ports	**MIB**	**Geosmin**	**MIB**	**Geosmin**	**MIB**	**Geosmin**
Tap Water	10.1	<2.0	7.0	9.0	5.2	3.4
After Aeration Tank	12.2	8.2	7.0	6.0	5.6	3.1
Flowing Water (PVC)	8.5	4.0	11.2	3.0	7.1	2.4
Flowing Water (cast-iron)	6.3	<2.0	4.7	<2.0	6.5	<2.0
Sampling Time	07/22/04		07/29/04		08/05/04	
Sampling Ports	**MIB**	**Geosmin**	**MIB**	**Geosmin**	**MIB**	**Geosmin**
Tap Water	7.2	6.8	4.8	3.1	14.4	4.4
After Aeration Tank	9.8	6.4	5.7	2.9	11.9	4.4
Flowing Water (PVC)	5.9	2.5	5.4	<2.0	7.3	<2.0
Flowing Water (cast-iron)	5.5	<2.0	5.2	<2.0	9.2	<2.0

Table 7.3
MIB and Geosmin levels in the tap water (June 6 – July 1, 2004)

Date	**MIB (ng/L)**	**Geosmin (ng/L)**
24-Jun	4.1	<2.0
25-Jun	8.6	2.3
26-Jun	9.5	2.2
27-Jun	3.1	<2.0
28-Jun	4.4	<2.0
29-Jun	N/A	N/A
30-Jun	6.5	6.8
1-Jul	10.1	<2.0

Table 7.4
Concentrations of actinomycetes and MIB in cast-iron pipeloop system

		Initial conc.		Before 12 days stagnation		After 12 days Stagnation
Sampling Port				Before drain	After drain	
Flowing water	CFU/mL	5.85E+04	8.10E+03	4.40E+03	1.00E+03	9.60E+04
	MIB(ng/mL)	6.2	5.3	4.5	4.3	33.5
Dead-end	CFU/mL	1.21E+05	2.46E+03	9.80E+03	1.05E+03	7.08E+04
	MIB(ng/mL)	58.4	4.5	4.3	5.1	36.5
Reservoir	MIB(ng/mL)	5.9	4.0	4.6	4.6	43.2
Biofilm	CFU/mm^2		7.625E+03			3.9E+2

Table 7.5.
Concentrations of actinomycetes and MIB in PVC pipeloop system

		Initial Conc.		Before 12 days stagnation		After 12 days Stagnation
Sampling Port				before drain	After drain	
Flowing Water	CFU/mL	2.5E+05	3.0E+04	1.7E+04	1.7E+04	7.0E+04
	MIB(ng/mL)	20.5	18.3	16.5	16	44.5
Dead-end	CFU/mL	2.7E+05	3.8E+04	1.6E+04	2.1E+04	7.3E+04
	MIB(ng/mL)	62.7	16.6	16.9	17.3	43.9
Reservoir	MIB(ng/mL)	20.7	16.3	16.8	16.8	59.5
Biofilm	CFU/mm^2		1.4E+1			1.8E+2

Table 7.6
Total and free chlorine concentrations in tap water during summer of 2004

	Tap Water		**Open Tank**	
Date	**Total** (mg/L Cl_2)	**Free** (mg/L Cl_2)	**Total** (mg/L Cl_2)	**Free** (mg/L Cl_2)
06/21/04	0.12	0.03	0.03	0.01
06/22/04	0.24	0.10	0.03	0.01
06/24/04	0.37	0.24	0.10	0.02
06/25/04	0.13	0.04	0.03	0.01
06/26/04	0.16	0.07	0.02	0.01
06/27/04	0.12	0.05	0.03	0.01
06/28/04	0.08	0.03	0.02	0.01
06/29/04	0.31	0.10	0.04	0.02
06/30/04	0.90	0.70	0.09	0.03
07/01/04	0.79	0.64	0.11	0.04
07/12/04	0.58	0.46	0.06	0.02
07/13/04	0.79	0.70	0.08	0.02
07/14/04	0.55	0.41	0.06	0.03

Table 7.7
Actinomycetes and MIB concentrations in cast-iron during the first chlorination event

	Chlorination								Cl depletion	
Time (day)	Day1								Day2	Day3
Time (hrs)	0.5	1	2	3	6	12	18	24	48	72
CFU/mL	57	58	26	54	20	28	105	174	1300	2380
MIB (ng/L)	2.3	3.6	2.0	2.0	2.6	2.6	2.3	2.1	2.5	2.1
Free Cl_2(mg/L)	0.76	0.96	0.78	1.15	1.01	1.09	1.01	1.02	0.00	0.00

Table 7.8
Concentrations of actinomycetes and MIB in cast-iron during second chlorination event (0.5 mg/L)

	Rechlorination										Cl depletion	
Time (day)	Day4					Day5					Day6	Day7
Time (hrs)	72.5	73	74	75	78	84	90	96	108	120	144	168
CFU/mL	144	58	163	36	62	127	192	223	51	233	2380	950
MIB (ng/L)	2.1	2.4	2.4	2.5	22.7	21.1	23.8	23.3	24.1	13.4	10.1	14.2
Free Cl_2(mg/L)	0.45	0.33	0.50	0.49	0.48	0.51	0.49	0.36	0.38	0.44	0.00	0.00

Table 7.9
Concentrations of actinomycetes and MIB in cast-iron during third chlorination event (2.0mg/L)

	Rechlorination							Cl depletion		
Time (day)	Day8							Day9	Day10	Day11
Time (hrs)	168.5	169	170	171	174	180	192	216	240	264
CFU/mL	50	30	9	35	24	27	17	528	4600	1250
MIB (ng/L)		16.3	18.5	19.6	14.0	15.7	16.3	16.2	15.5	14.1
Free Cl_2(mg/L)	1.45	1.91	1.61	1.39	1.56	1.71	1.53	0.00	0.00	0.00

Table 7.10
The concentrations of actinomycetes and MIB in PVC during the first event chlorination (1.0mg/L)

	Chlorination								Cl depletion	
Time (day)	Day1								Day2	Day3
Time (hrs)	0.5	1	2	3	6	12	18	24	48	72
CFU/mL	227	43	72	44	3	5	10	15	1440	2540
MIB (ng/L)	2.6	2.0	2.0	3.6	2.0	2.0	2.0	2.0	2.3	2
Free Cl_2(mg/L)	0.95	1.26	0.91	1.20	1.81	1.07	1.01	0.99	0.00	0.00

Table 7.11
Actinomycetes and MIB concentrations in PVC during the second chlorination event (0.5mg/L)

	Rechlorination										Cl Depletion	
Time (day)	Day 4					Day 5					Day6	Day7
Time (hrs)	72.5	73	74	75	78	84	90	96	108	120	144	168
CFU/mL	600	927	318	80	280	33	41	80	12	111	1620	4000
MIB (ng/L)	2	2	2.3	20.5	22.4	21.7	24.6	21.7	21.3	17.9	14.9	13.2
Free Cl_2(mg/L)	0.45	0.33	0.50	0.49	0.48	0.51	0.49	0.36	0.38	0.44	0.00	0.00

Table 7.12
Actinomycetes and MIB concentrations in PVC during the third chlorination event (2.0mg/L)

	Rechlorination							Cl depletion		
Time (day)	Day8							Day9	Day10	Day11
Time (hrs)	168.5	169	170	171	174	180	192	216	240	264
CFU/mL	158	16	30	65	21	33	39	640	1008	3640
MIB (ng/L)		17.7	19.4	17.6	19.2	16.7	14.8	15.4	13.2	15.2
Free Cl_2(mg/L)	1.72	2.05	1.91	1.86	2.00	1.98	1.79	0.00	0.00	0.00

Table 7.13
Actinomycetes populations and MIB/Geosmin concentrations in cast-iron coupons recovered from the city's distribution system

Coupon No.	**1**	**2**	**3**
Area (cm^2)	120.8	124.7	56.13
CFU/mL	1.62E+06	3.15E+06	1.37E+05
(CFU/mL) x 50mL	8.10E+07	1.58E+08	6.85E+06
CFU/cm^2	6.71E+05	1.26E+06	1.22E+05
MIB (ng/L)	2.8	<2.0	50.6
Geosmin (ng/L)	<2.0	3.6	2.6

Table 7.14
Actinomycetes and MIB/Geosmin concentrations in water samples received with coupons recovered from the city's distribution system

Coupon No.	**1**	**2**	**3**
CFU/mL	1.45E+05	4.41E+05	4.41E+05
MIB (ng/L)	5.8	12.2	12.2
Geosmin (ng/L)	<2.0	<2.0	<2.0

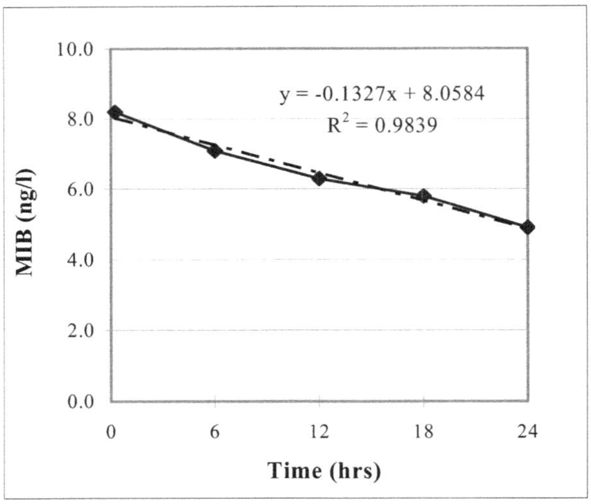

Figure 7.1 Depletion rate of MIB at low concentrations in cast-iron during initial 24 hours

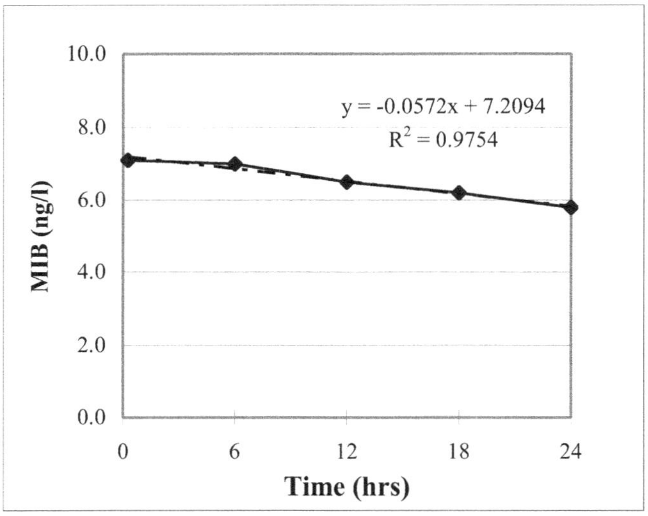

Figure 7.2 Depletion rate of MIB at low concentrations in PVC during initial 24 hours

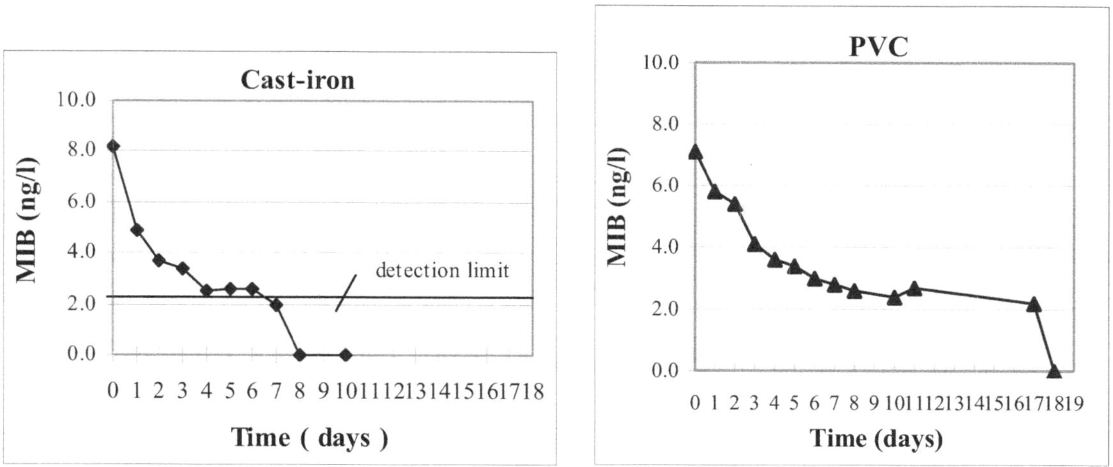

Figure 7.3 Depletion rate of MIB at low concentrations in cast-iron and PVC

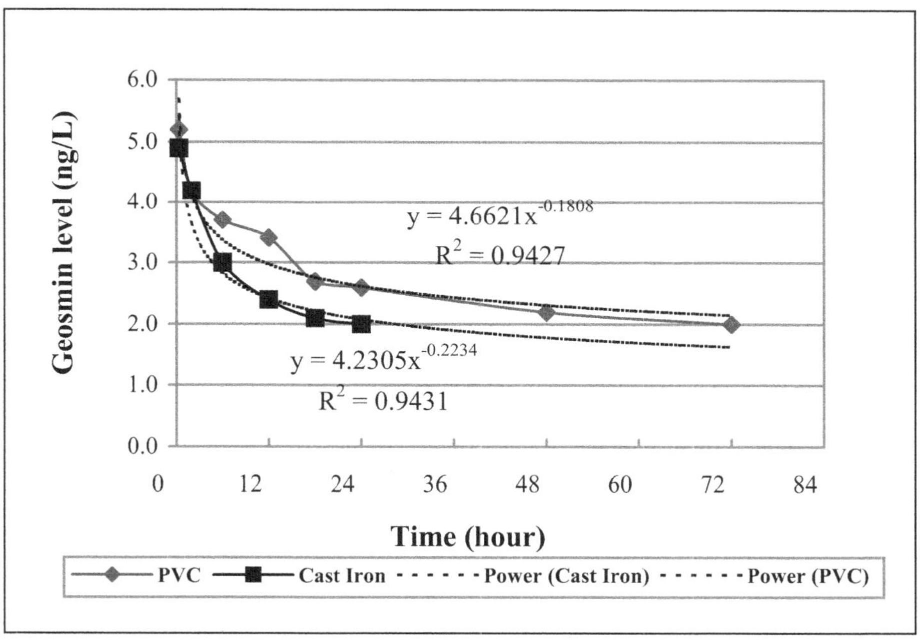

Figure 7.4 Depletion rate of Geosmin in cast-iron and PVC

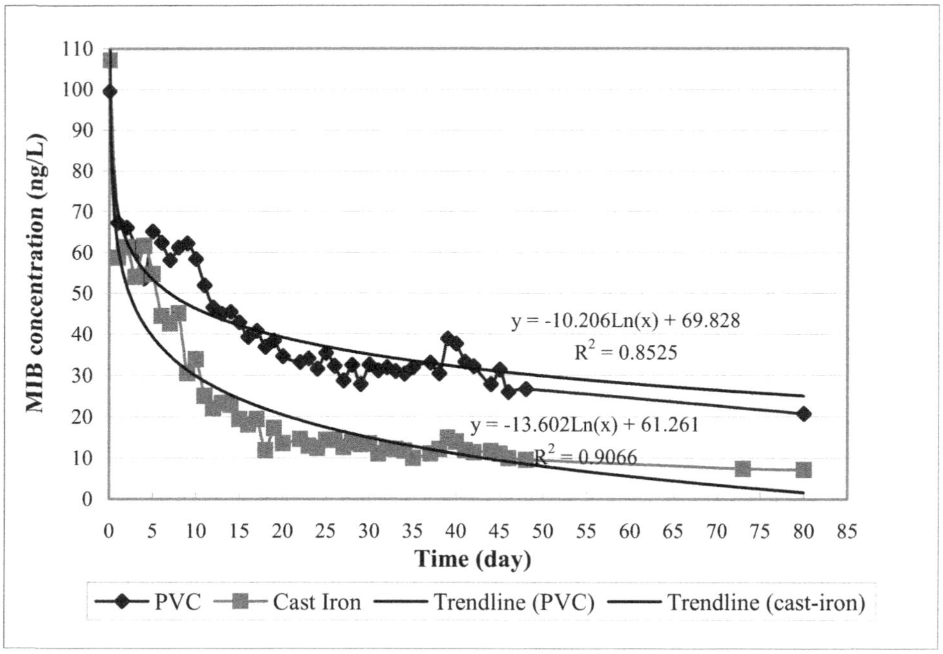

Figure 7.5 Depletion rate of MIB at high concentrations in PVC and cast-iron

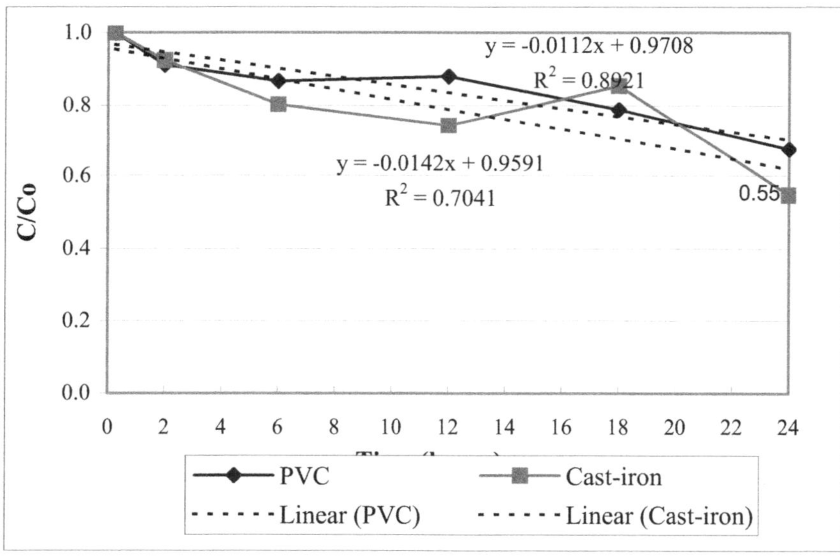

Figure 7.6 Depletion rate of MIB during the initial 24 hours in PVC and Cast-iron

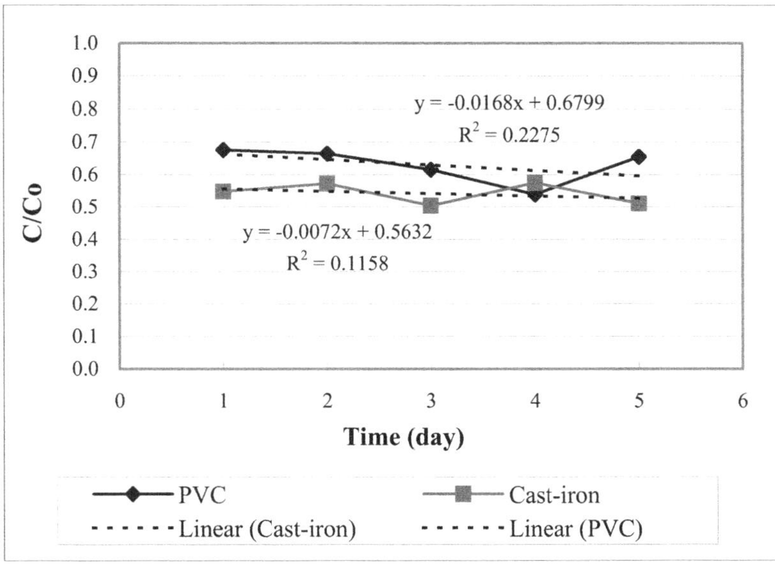

Figure 7.7 Depletion rate of MIB during day 1 to day 5

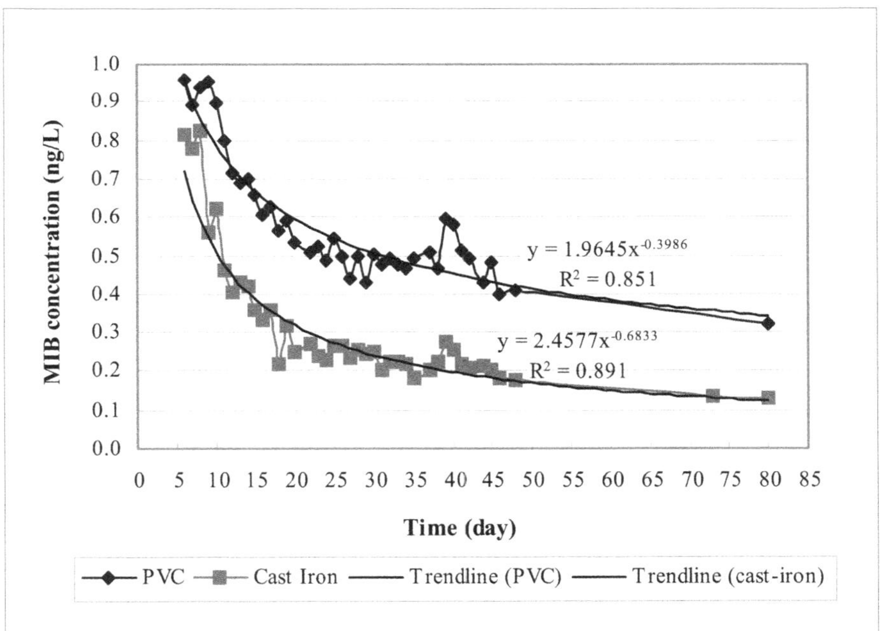

Figure 7.8 Depletion rate of MIB after day 6 till the end of experiment

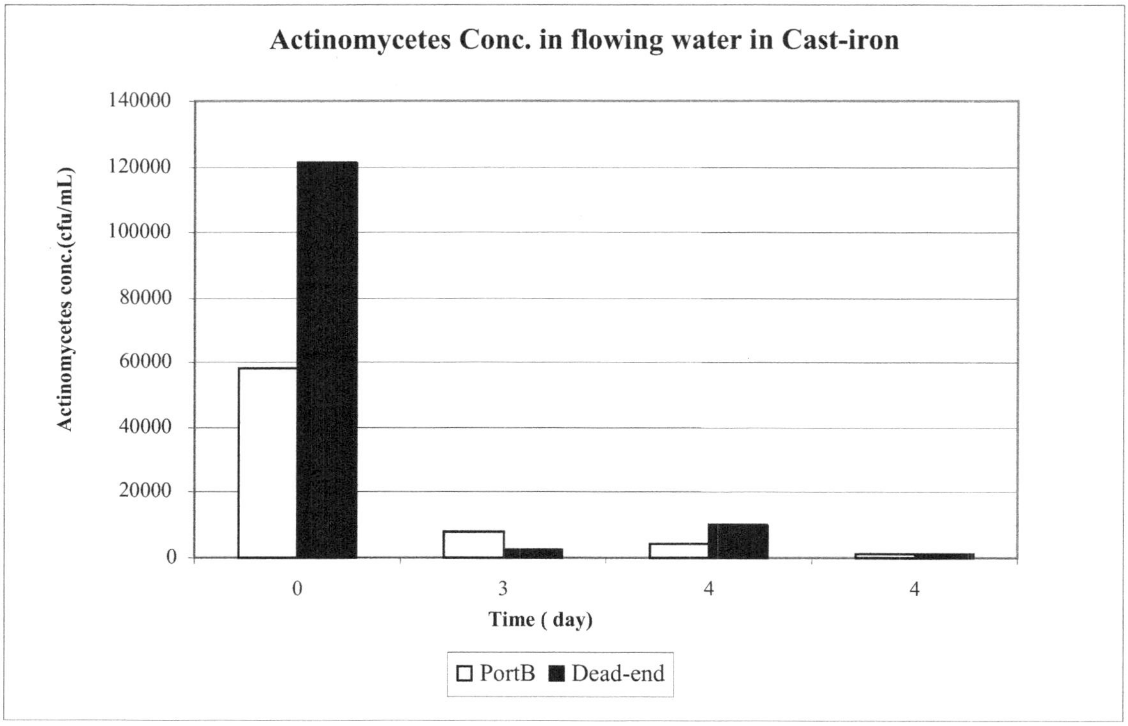

Figure 7.9 Concentrations of actinomycetes in flowing water in cast-iron during the stabilization and stagnation period

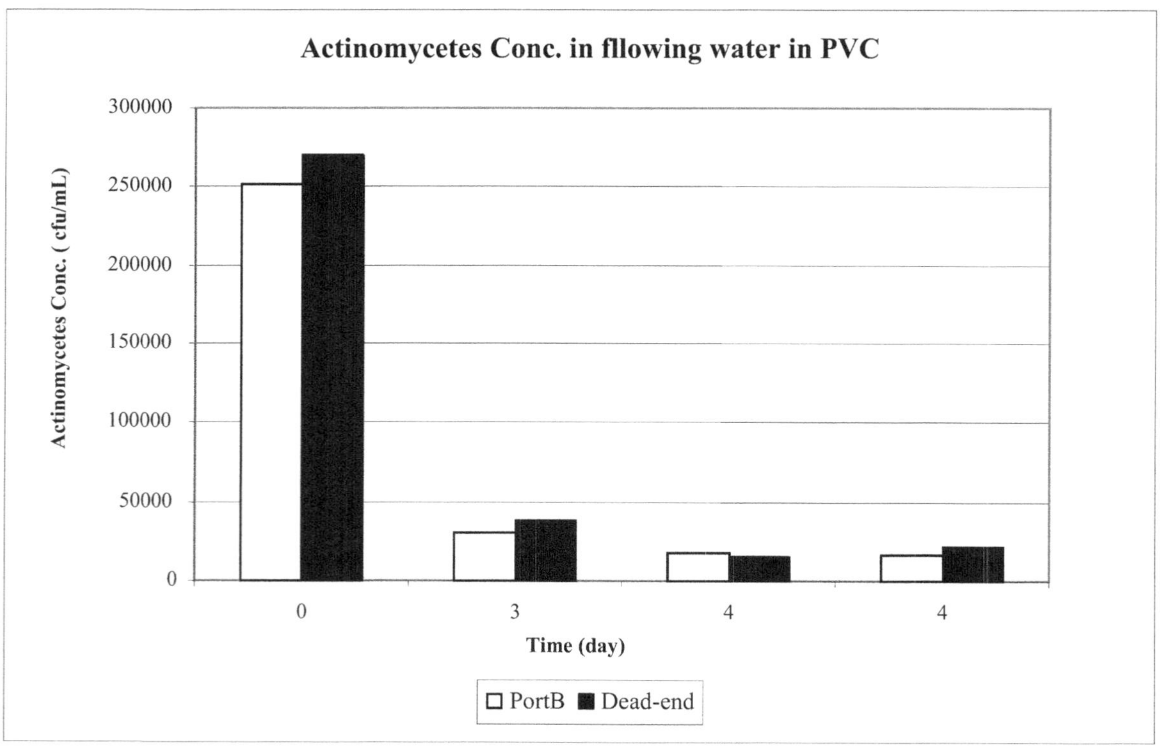

Figure 7.10 Concentrations of actinomycetes in flowing water in PVC during stabilization and stagnation period

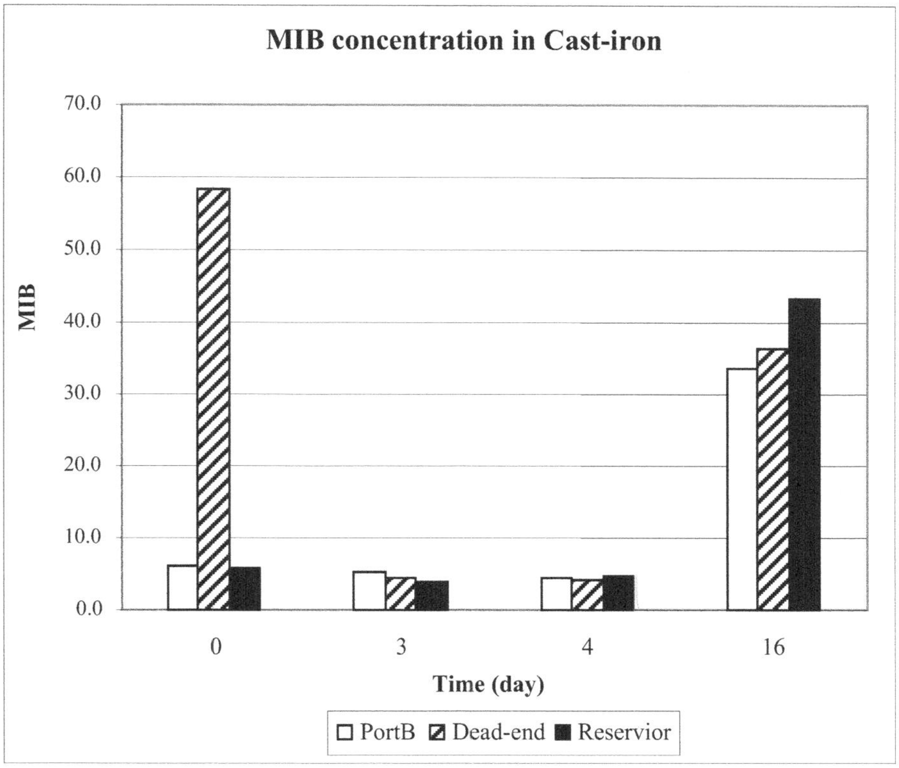

Figure 7.11 MIB concentration in flowing water in cast-iron pipeloop over time

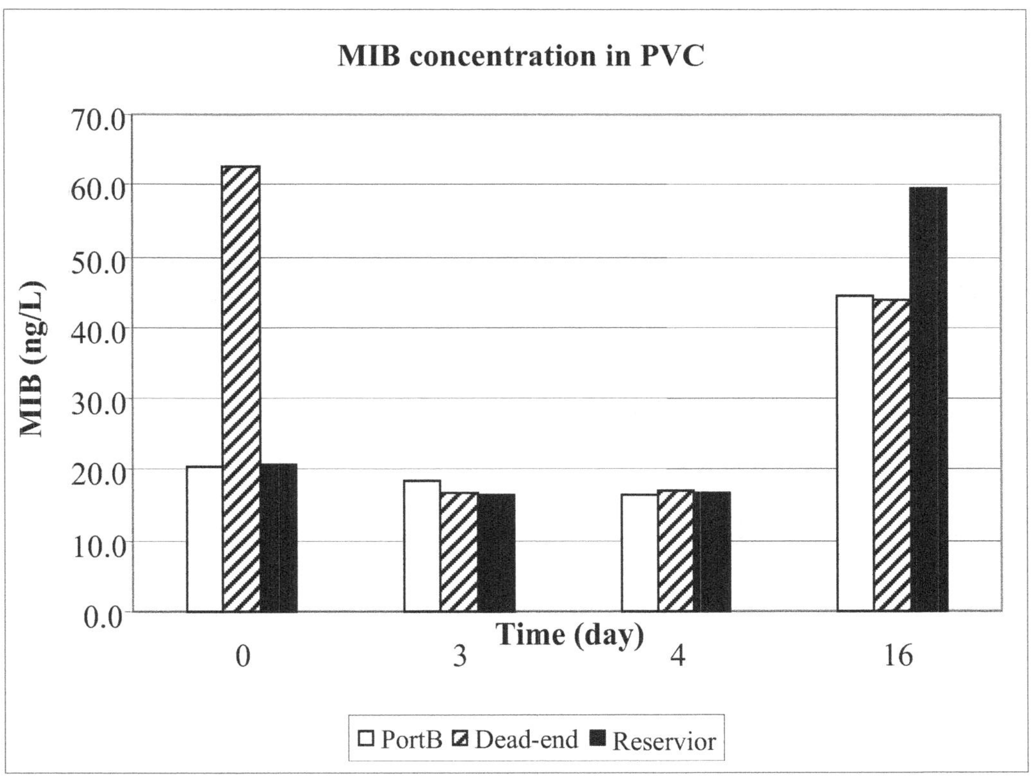

Figure 7.12 MIB concentration in PVC pipeloop in flowing water over time.

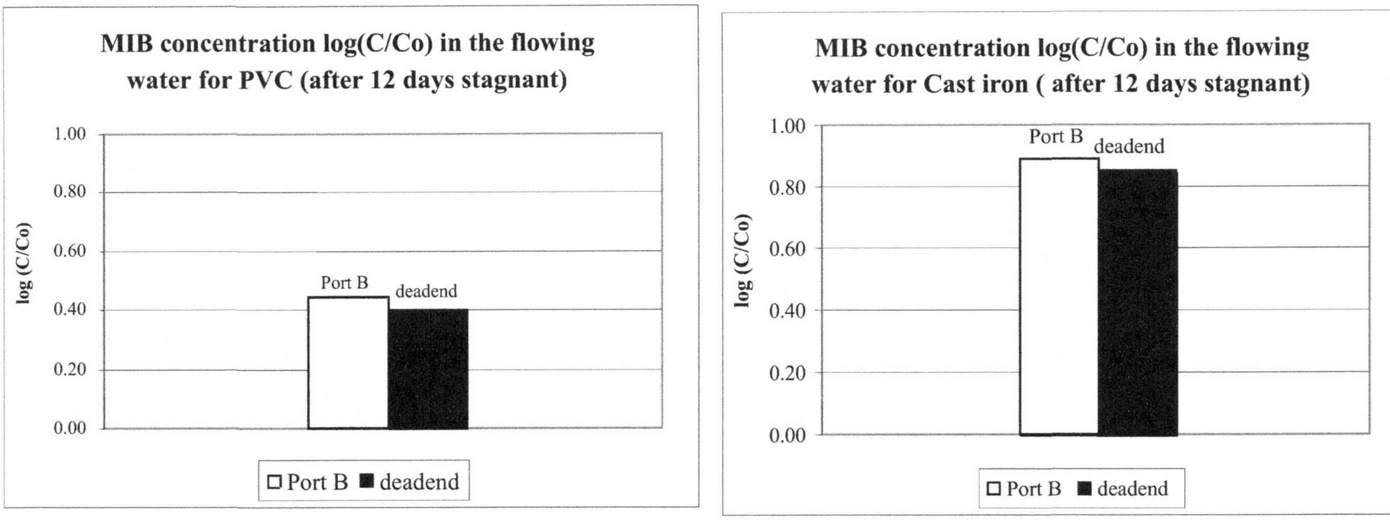

Figure 7.13 Effect of stagnation on MIB concentrations in flowing water

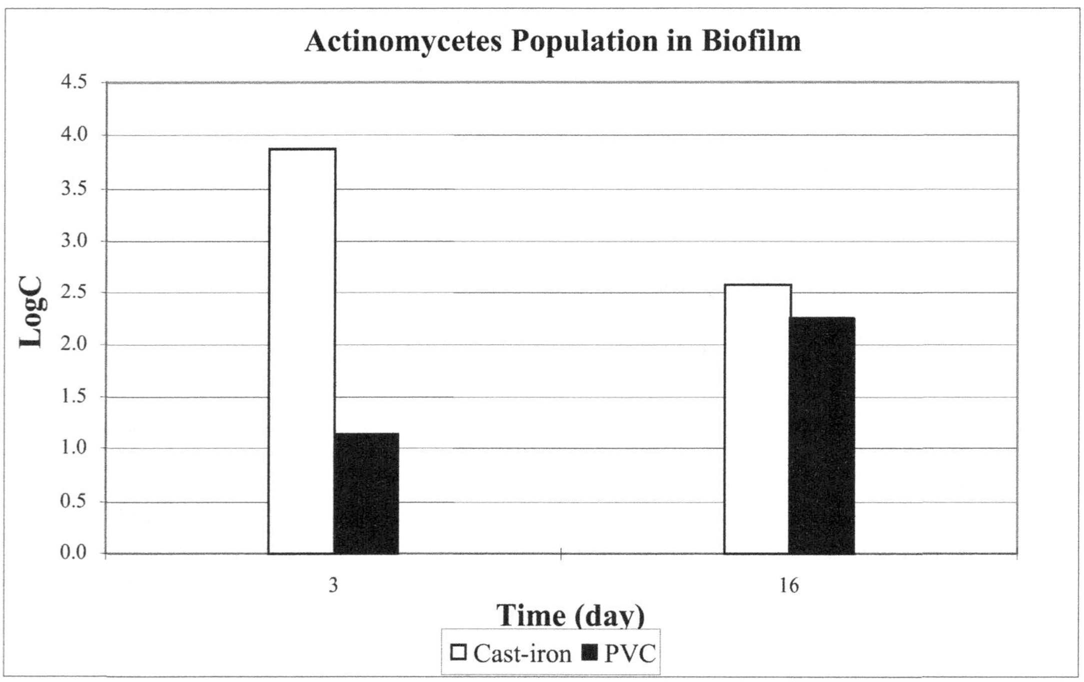

Figure 7.14 Effect of stagnation on actinomycetes population in biofilms

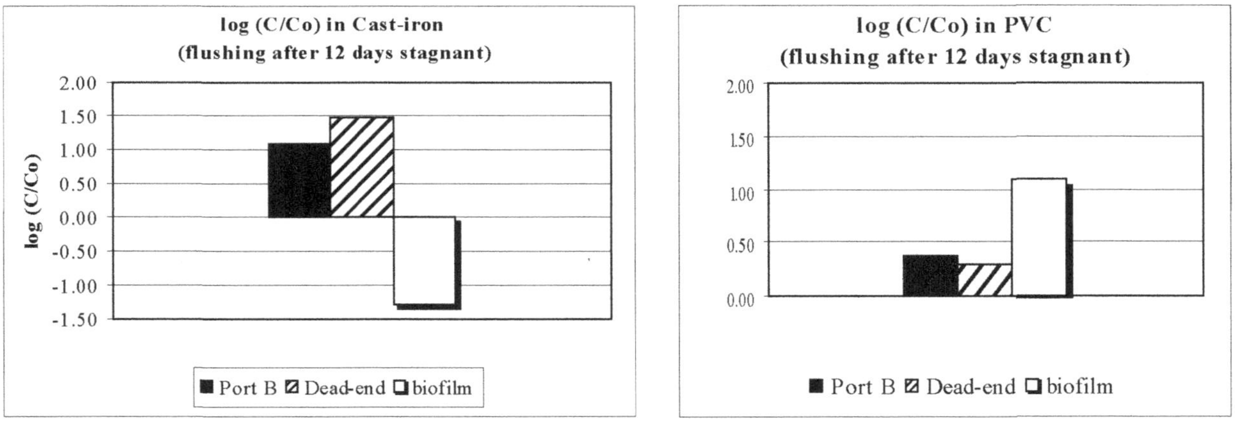

Figure 7.15 Effect of stagnation on actinomycetes population in flowing water and biofilm

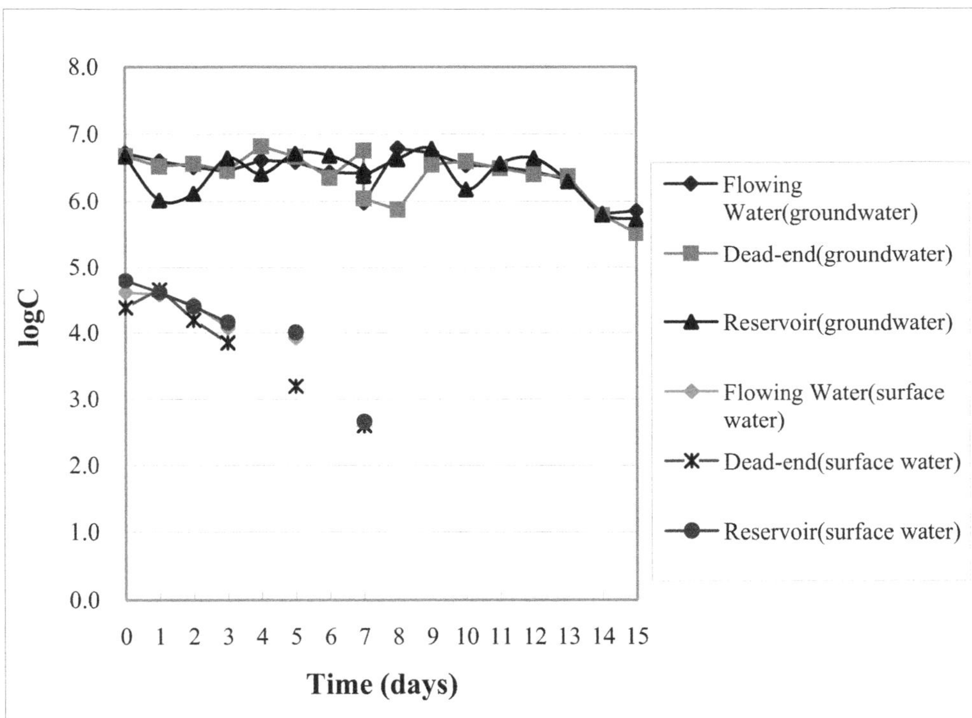

Figure 7.16 Concentrations of actinomycetes in groundwater and surface water in PVC

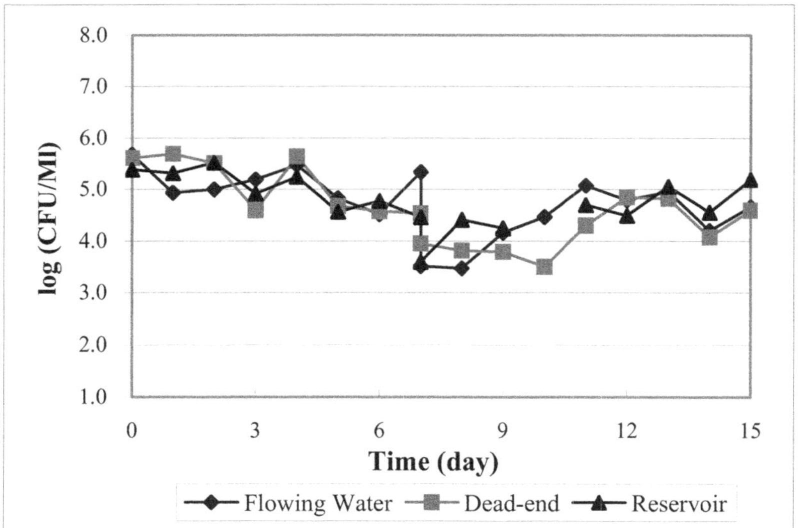

Figure 7.17 Concentrations of actinomycetes in cast-iron supplied with groundwater

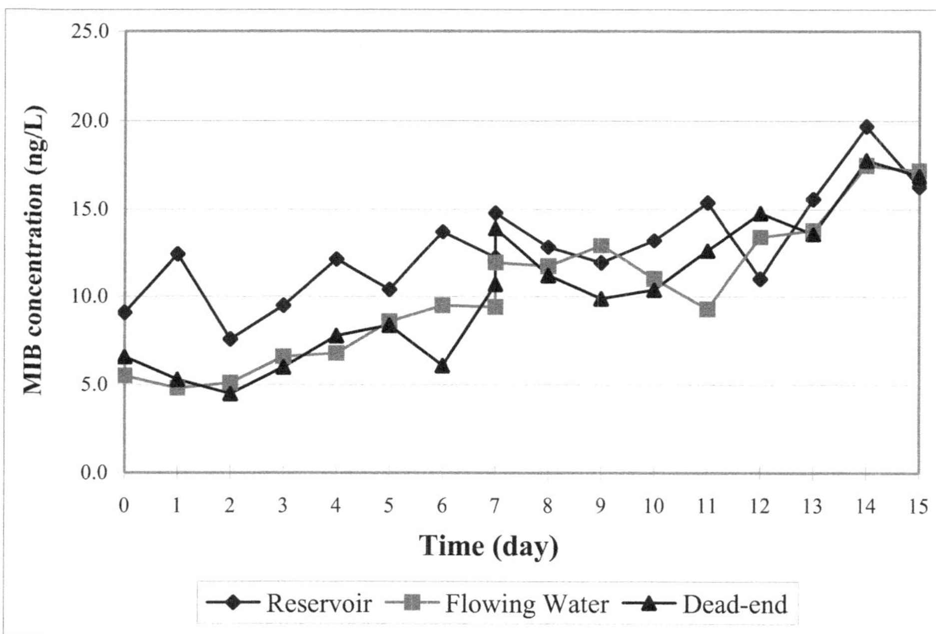

Figure 7.18 Concentrations of MIB in PVC supplied with groundwater

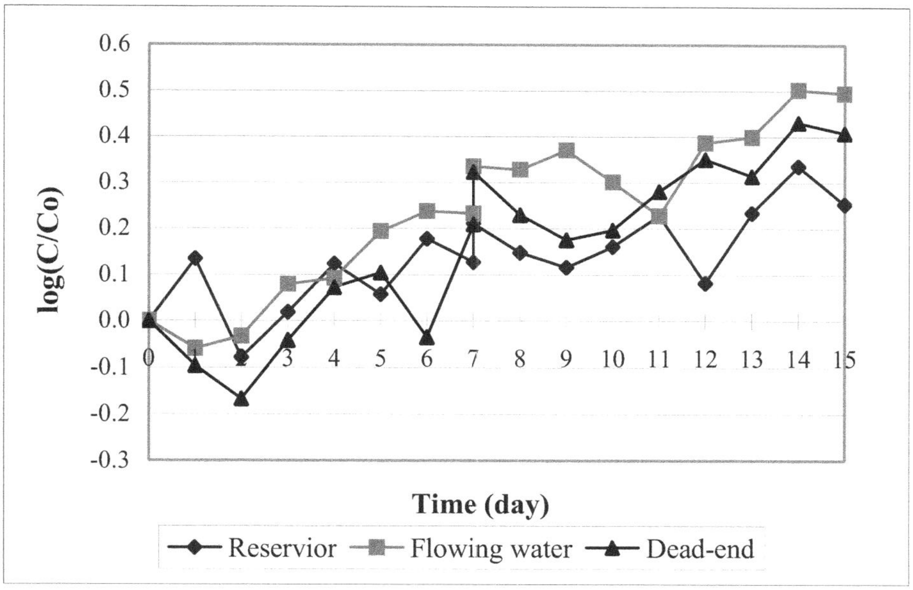

Figure 7.19 Concentrations of MIB (log C/Co) in PVC supplied with groundwater

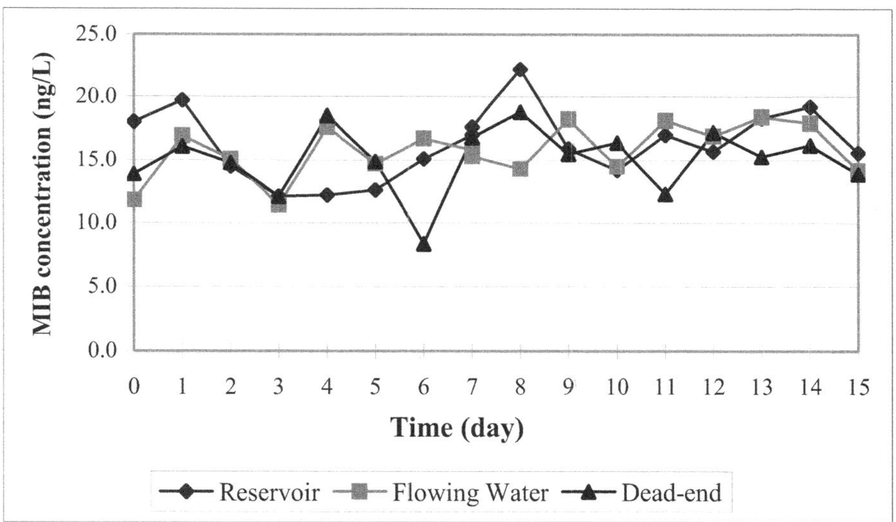

Figure 7.20 Concentrations of MIB in groundwater in cast-iron

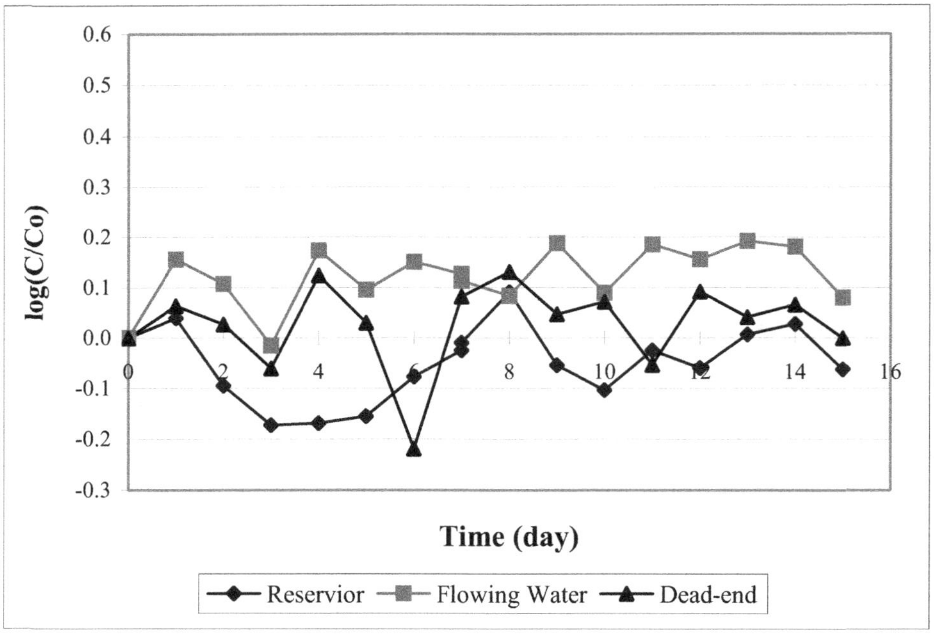

Figure 7.21 Concentrations of MIB in groundwater in cast-iron

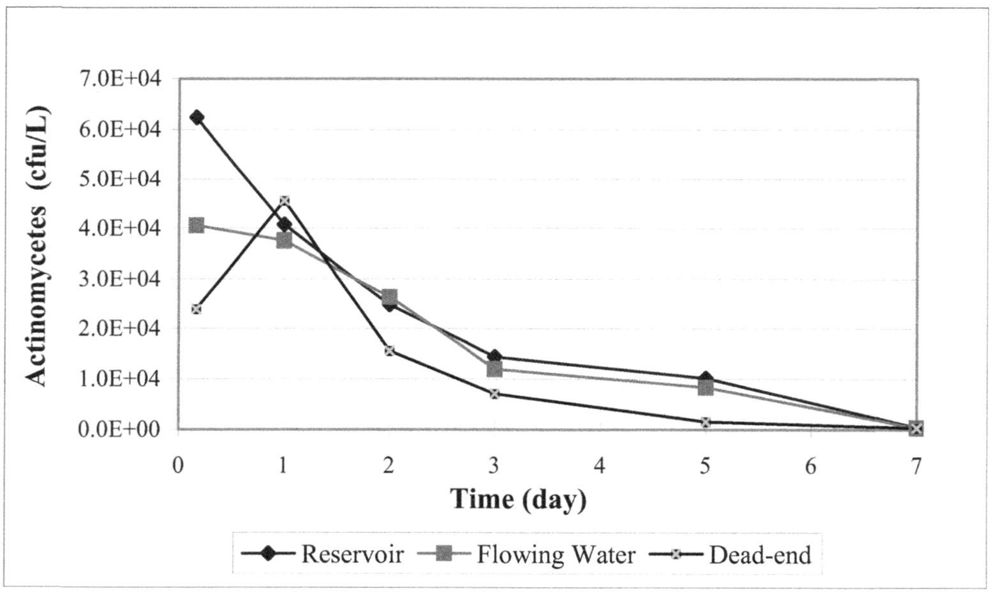

Figure 7.22 Concentration of actinomycetes in surface water in PVC pipeloop

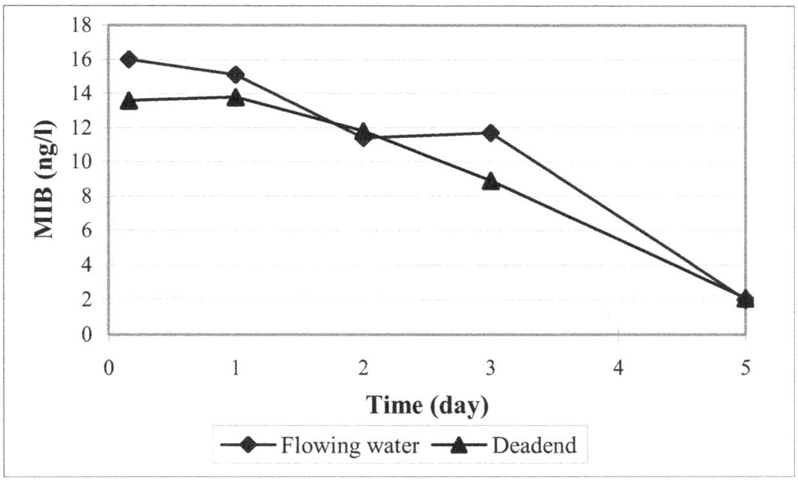

Figure 7.23 MIB concentrations in surface water in PVC pipeloop.

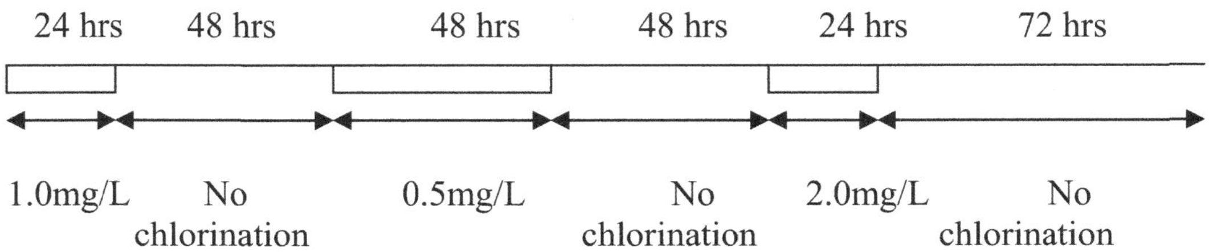

Figure 7.24 Time line of chlorination and rechlorination experiment

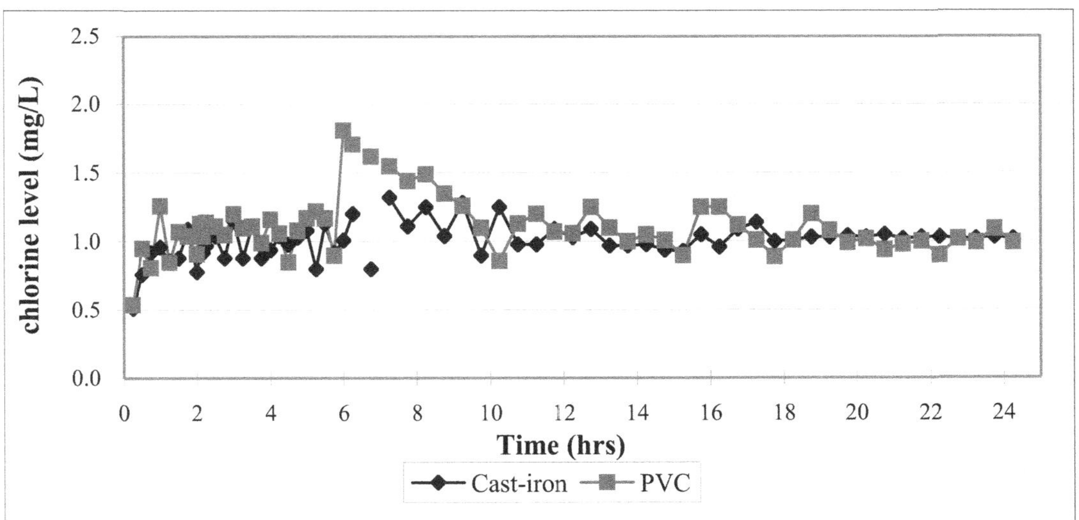

Figure 7.25 Free chlorine residual in PVC and cast-iron during the first chlorination-episode (0 - 24 hr)

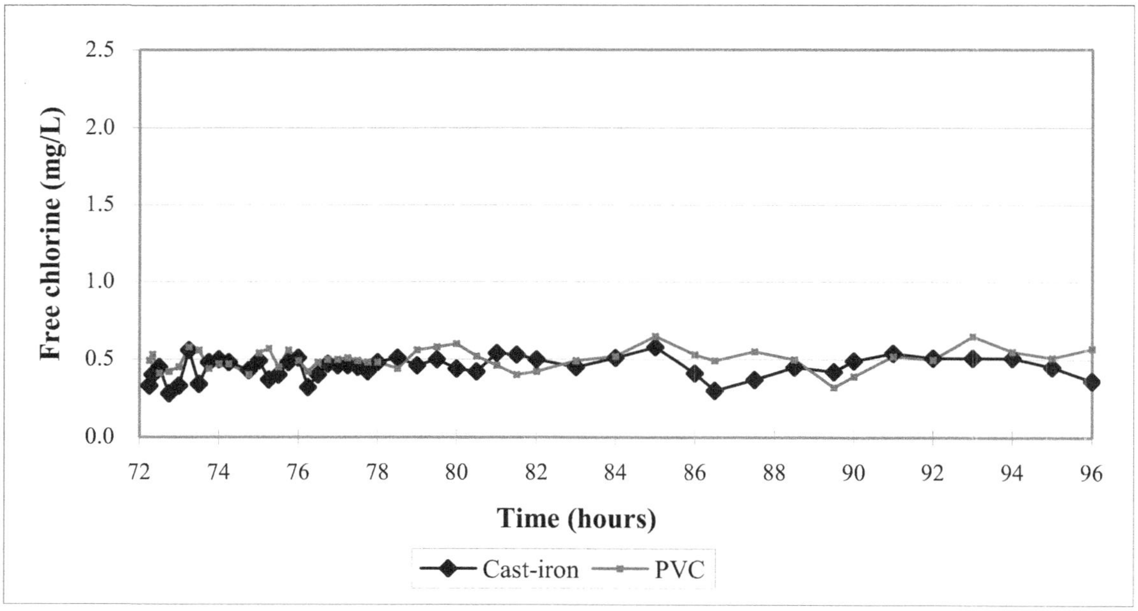

Figure 7.26 Free chlorine residual in cast-iron and PVC during the first day of the second chlorination-episode (72 - 96 hr))

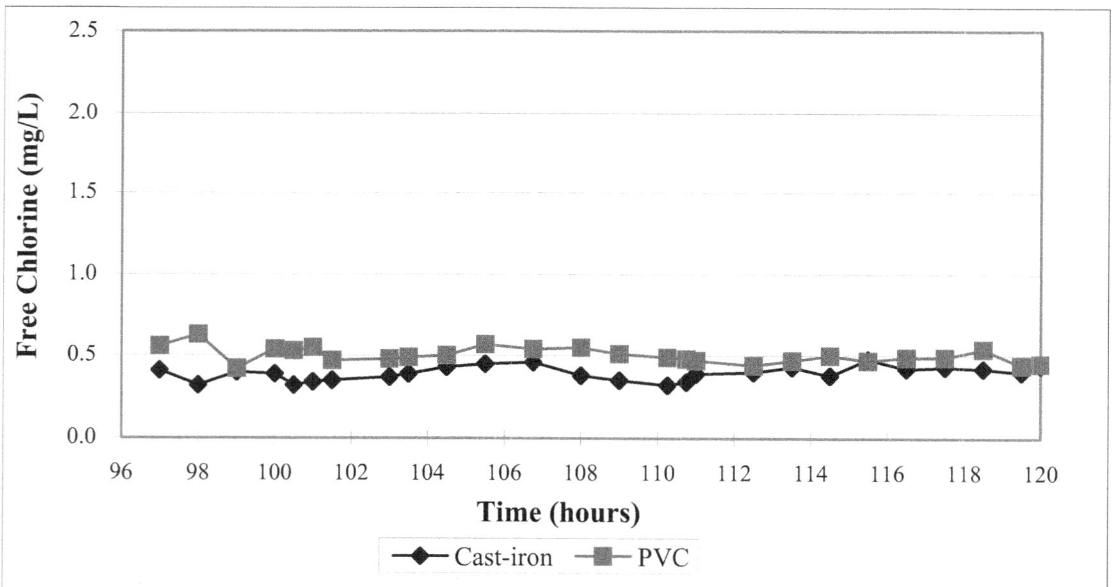

Figure 7.27 Free chlorine residual in PVC and cast-iron during the second day of the second chlorination-episode (96 - 120 hr)

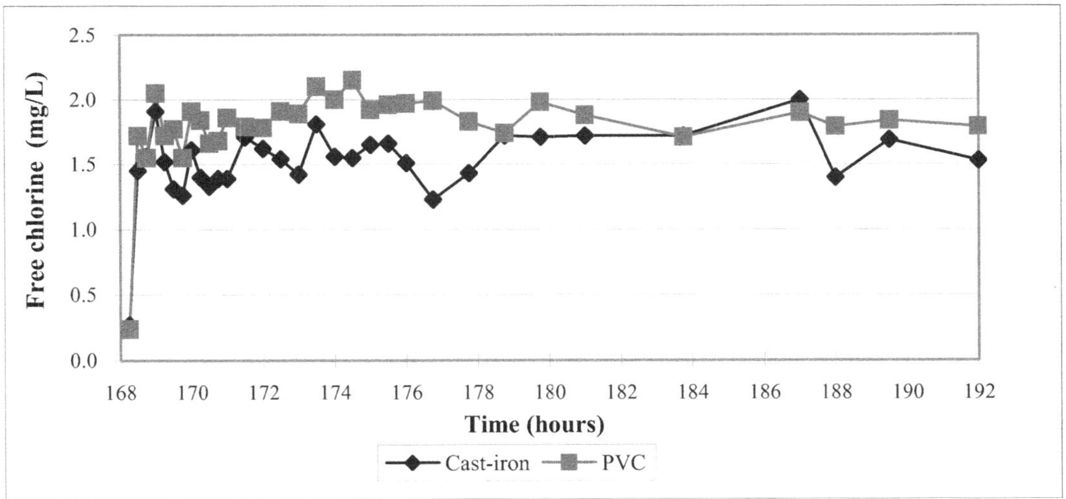

Figure 7.28 Free chlorine residual in PVC and Cast-iron during the third chlorination-episode (168 ~ 192 hr)

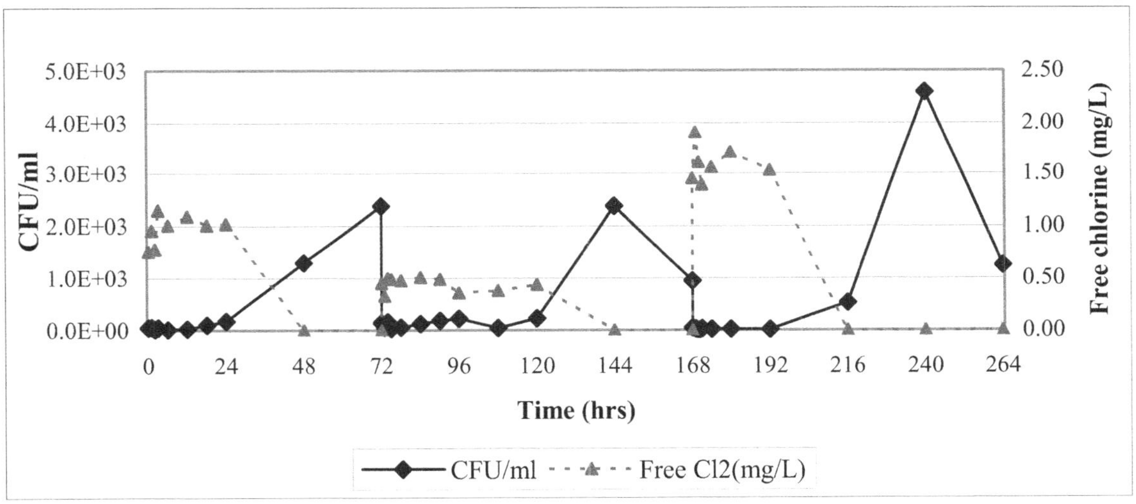

Figure 7.29 Concentrations of actinomycetes in flowing water in cast-iron during chlorination and rechlorination

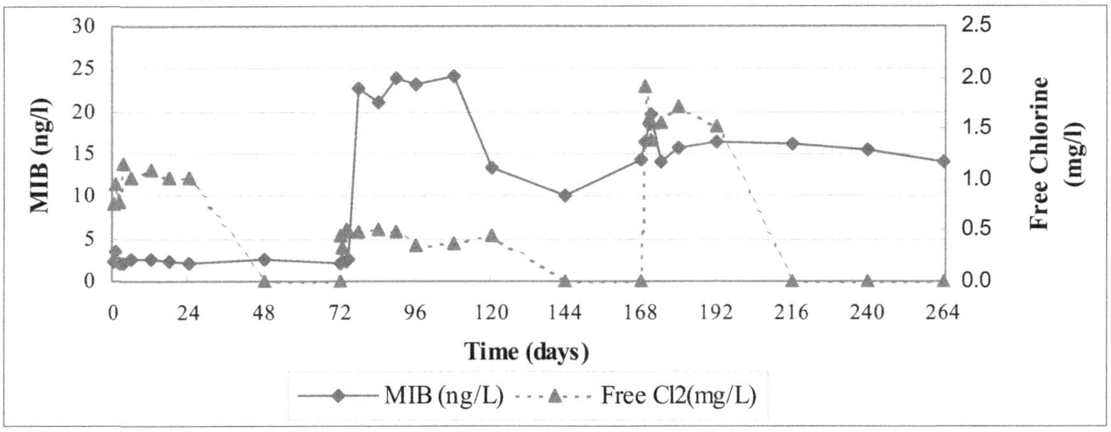

Figure 7.30 Concentrations of MIB in flowing water in cast-iron during chlorination and rechlorination

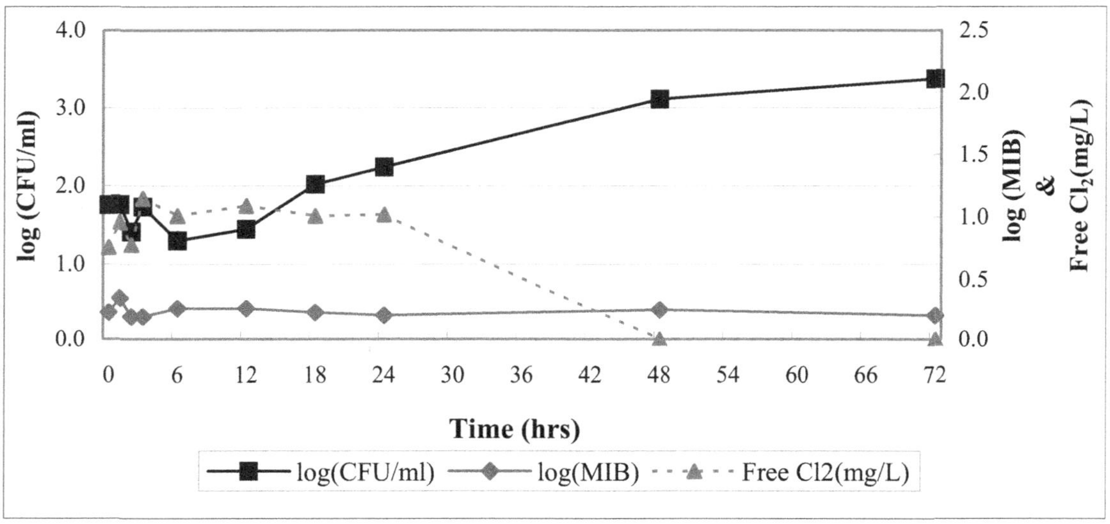

Figure 7.31 Actinomycetes and MIB concentrations in cast-iron and PVC during the first chlorination event (1.0 mg/L)

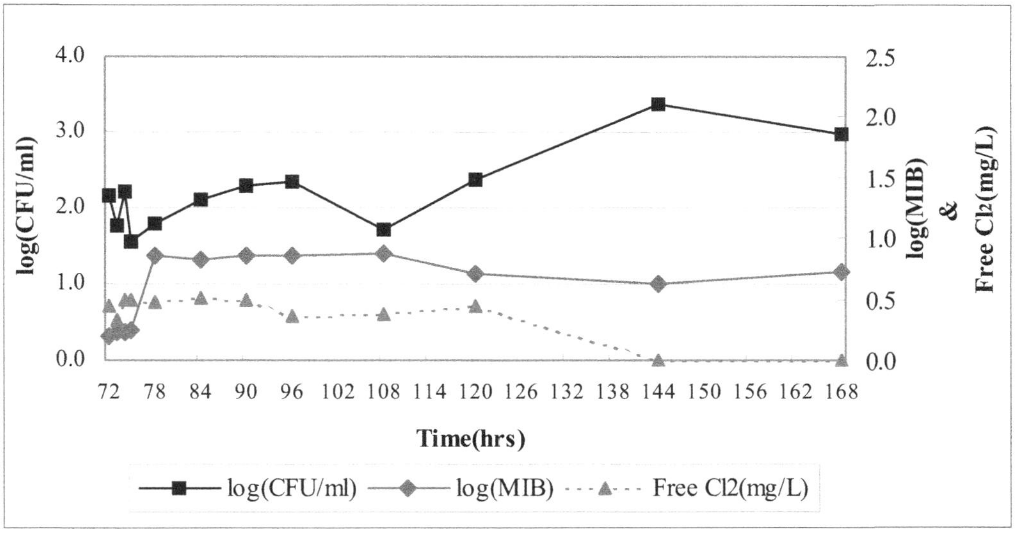

Figure 7.32 Actinomycetes and MIB concentrations in cast-iron during the second chlorination event (0.5 mg/L)

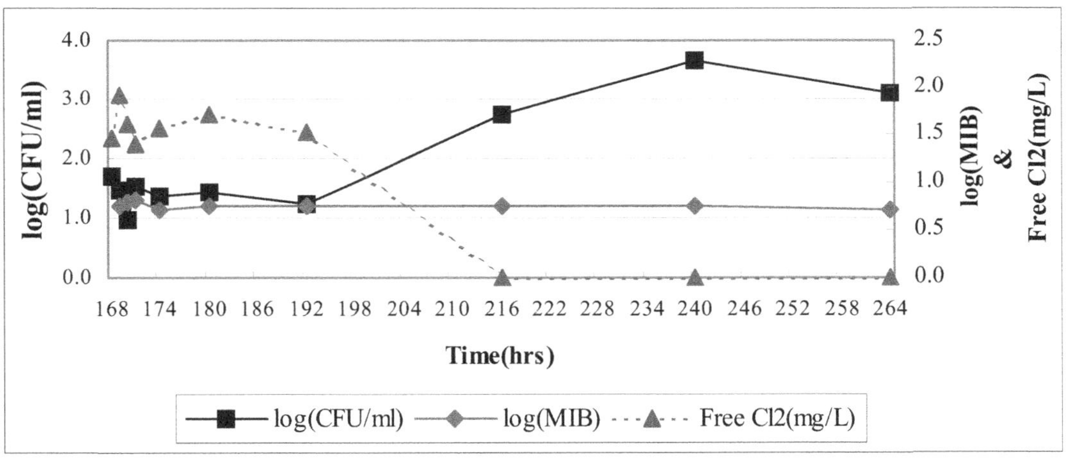

Figure 7.33 Actinomycetes and MIB concentrations in cast-iron during the third event of chlorination (2.0 mg/L)

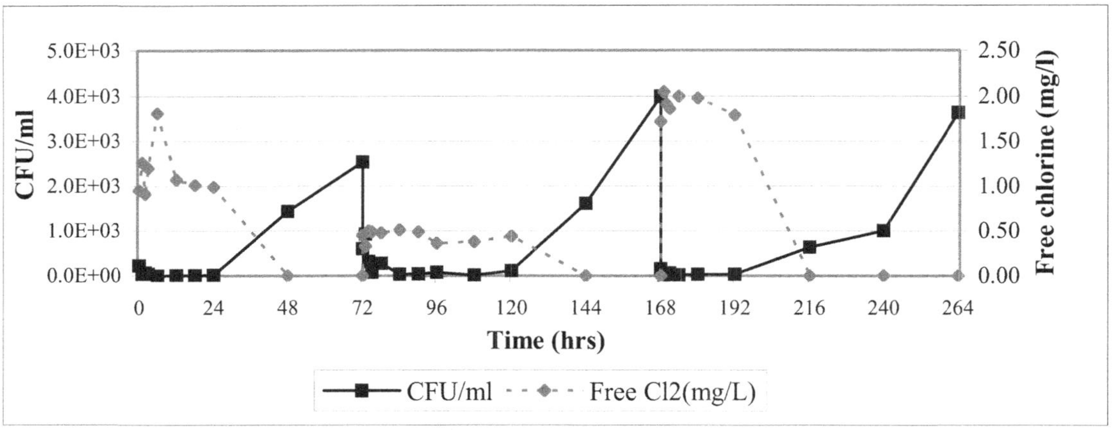

Figure 7.34 Concentrations of actinomycetes in flowing water in PVC during chlorination and rechlorination

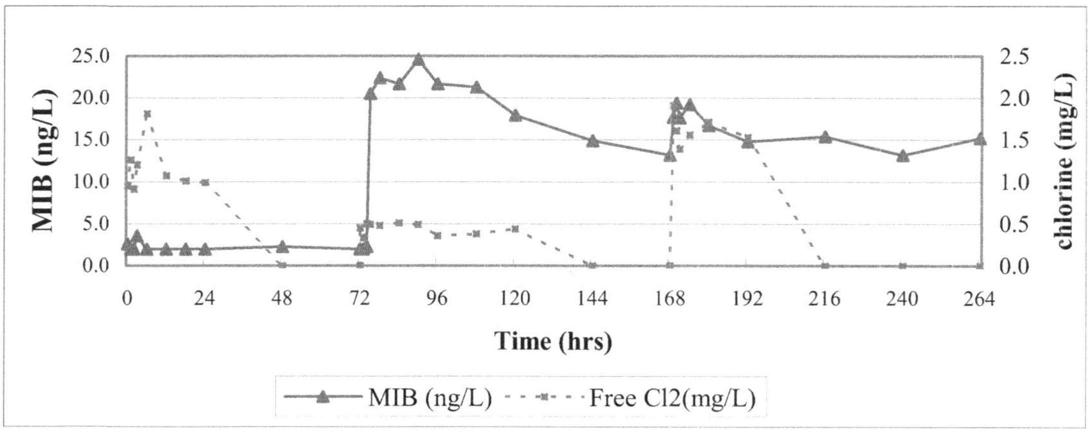

Figure 7.35 Concentrations of MIB in flowing water in PVC during chlorination and rechlorination

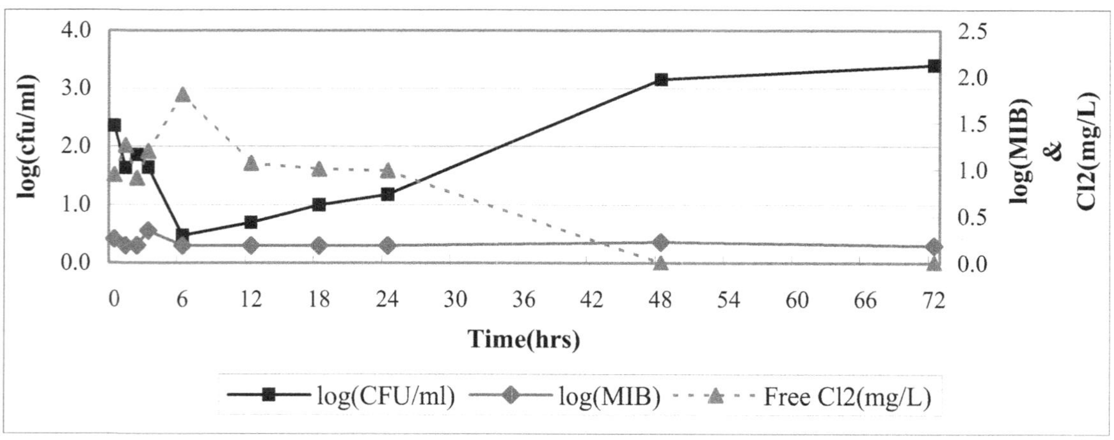

Figure 7.36 Actinomycetes and MIB concentrations in PVC during the first chlorination event (1.0 mg/L)

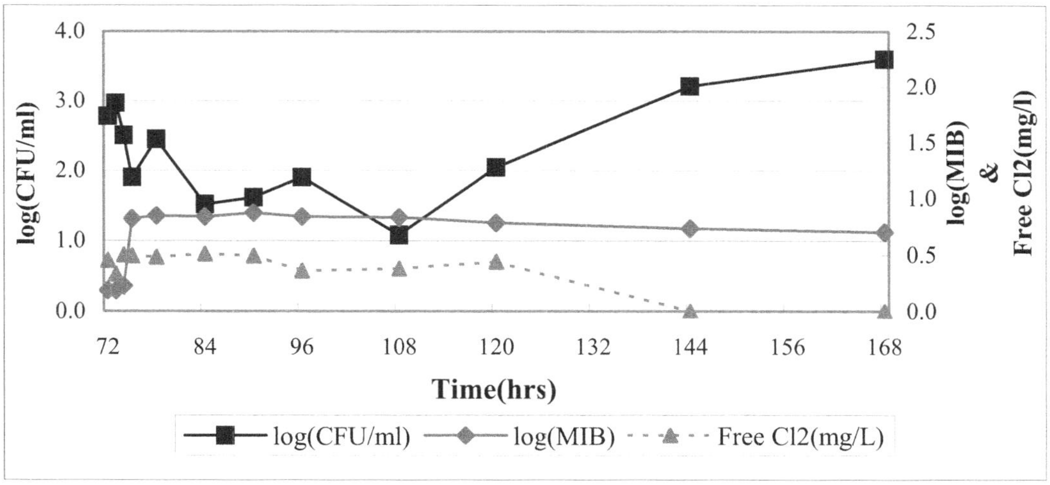

Figure 7.37 Actinomycetes and MIB concentrations in PVC during the second chlorination event (0.5 mg/L).

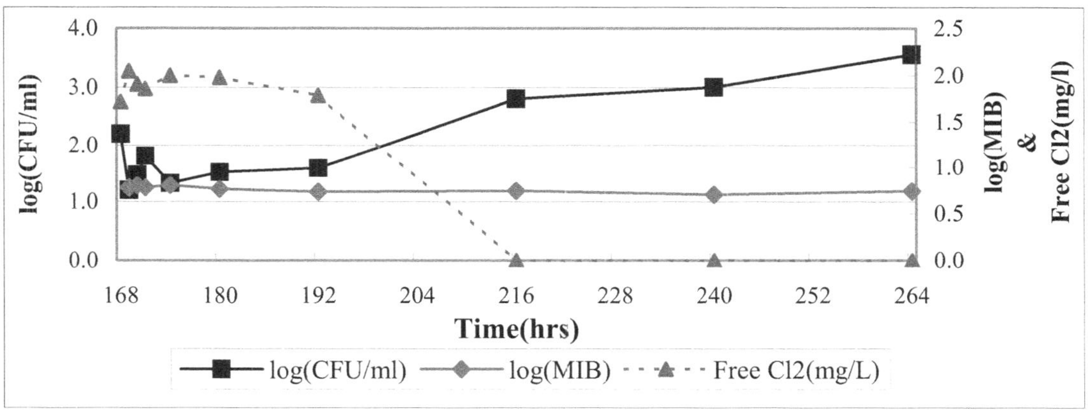

Figure 7.38 Actinomycetes and MIB concentrations in PVC during the third chlorination event (2.0 mg/L)

CHAPTER 8
FINDINGS AND RECOMMENDATIONS TO UTILITIES AND FUTURE RESEARCH

FINDINGS

Development and effectiveness of a Flavor Profile Analysis (FPA) group

1. Increase in chlorine masked odors and fatigued panel members. Dechlorination of monitoring samples with hydrogen peroxide is recommended.
2. Earthy/Musty FPA showed a direct correlation with MIB and Geosmin but a weak correlation with β-cyclocitral.
3. Earthy/Musty FPA intensity could be predicted with MIB and Geosmin plots and with a three component (MIB, Geosmin, β-cyclocitral) plot.
4. FPA panel staffing proved to be difficult due to time and priorities of the Water Department. Training new members was continuous. SPME GC/MS analysis is recommended for fast and accurate results.

Quantification of Non-MIB/Geosmin Taste and Odors

1. The SPME GC/MS method was expanded to include cis-3-hexenen[1]-ol, cis-3-hexenyl acetate, 2-isobutyl-3-methoxypyrazine (IBMP), 2-isopropyl-3-methoxypyrazine (IPMP), and β-cyclocitral.

Microcystin Method Validation and Comparison

1. The ELISA assays conducted were considered acceptable based on their high R^2 values. ELISA assays are recommended for future MC analyses.
2. The PP2A assay was unaffected by water matrices, and 74% of the assays conducted were considered acceptable. Assays were considered unacceptable based on inconsistent calibration results, lack of color production, or poor R^2 values (< 0.75). Enzyme activation at room temperature was considered the main source of error in PP2A assays. PP2A is not recommended for future MC monitoring.
3. Freezing of samples before MC analysis affects measured MC concentrations, and further research should be conducted to determine the source of this concentration change.
4. ELISA and PP2A predicted similar concentrations for three ELISA kit standards.
5. SPE with C18 cartridges provides a means of measuring MC in the part per trillion range.
6. Method C (24 hours in MeOH) is recommended for future intracellular extraction based on relative ease and completeness of extraction.

Algae Metabolites in Raw and Treated Drinking Waters

1. Several cyanobacteria were identified and isolated as part of the City of Phoenix/ASU project. Potential MC producers include *Anabaena* and *Oscillatoria*.
2. MIB levels are higher in the upper layers of the reservoir (epilimnion).
3. MIB production in the canal system occurred noticeably only in the Arizona Canal during the warmer months of 2003.
4. β-cyclocitral can be produced by algae in the COP water supply and may play a significant role in the T&O occurrence.
5. July was the peak month for MC production with the highest dissolved concentration in Lake Pleasant.
6. PAC used for T&O compound absorption did reduce MC concentration.
7. No linear correlations ($R^2 > 0.5$) were found between MC concentration and chlorophyll-a, MIB, phosphorous, secchi disc depth, and POC.
8. MC peak production precedes MIB peak concentrations by one month.
9. MC and MIB are produced by different organisms. The presence of MIB does not indicate MC is present or absent. MC should be monitored continuously.

Water Distribution System Monitoring

1. Baseline distribution system monitoring did not show significant differences from MIB or Geosmin leaving the water treatment plants. Distribution system production of T&O did not occur or could not be detected.
2. Response to customer complaints must include classification of T&O (i.e. Earthy/Musty, Sewer, Fishy, Swampy).
3. Intensive distribution system monitoring did not show significant differences from MIB or Geosmin leaving the water treatment plants. The data do not indicate that production of T&O are based on time and distance from a plant.
4. Assumptions of water originations were incorrect for the intensive monitoring. It is recommended that intensive monitoring is repeated in different parts of the city.

Laboratory Studies of Actinomycete Inactivation

1. *Steptomycetes* inactivation at pH 5 occurred immediately after chlorine addition.

Biofilms in Water Distribution System

1. Despite attempts to obtain biofilms from the water distribution system, no good cultures were obtained. Well developed methods are available for flowing water but not for biofilms. Sampling protocols need to be developed for collection of biofilm samples.
2. MIB depletion rate in the pipeloop is faster in cast-iron than in PVC.
3. Stagnation followed by rapid water flux results in increased actinomycetes and MIB concentrations in flowing water.

4. Change in source water provides favorable conditions for the survival and production of actinomycetes and MIB in drinking water distribution systems.
5. Rechlorination of the pipeloop systems results in the release of high concentrations of MIB in flowing water. However, maintenance of chlorine at 2.0 mg/L had no impact on MIB concentration. Further investigation of MIB behavior in the distribution system is recommended.

FUTURE RESEARCH

Future research should include monitoring of MC through the water treatment plants in the Phoenix Metro area. Monitoring of MC in the reservoirs is recommended based on the dynamic nature of cyanobacteria blooms.

Additional research should be done to determine the effectiveness of WTP processes on MC removal.

APPENDIX A

CITY OF PHOENIX
WATER SERVICES DEPARTMENT
POLLUTION CONTROL DIVISION

Sample Log# ______________

COMPLAINT / INQUIRY FORM

Date Received: ___________ Inspector: _________ Time: _________________

Name of Consumer: ____

Address: ______________________________________ Zip Code: ___________________
(Ave., Ln., Dr. / St., Pl., Way)

Telephone: (_______)________________________________

Referred by ___________________ Follow-up Requested: Yes ☐ No ☐

If Yes, how? ____

Type of complaint: Information Request ☐ Illness ☐

Taste ☐ Odor ☐ Dirt/ Color/Sediment ☐ Chlorine ☐ Hardness ☐

Questions to Ask	
Hot Water or Cold Water?	Home Treatment System?
Dead End or Cul-de-sac?	How long has there been a problem?

Describe the complaint: ____

____

____

Comments (after call-back or visit): ____

____

____

____

____

____

Follow-up completed and case closed by:_____________________________Date:______________

REFERENCES

An, J. and W.W. Carmichael. 1994. *Use of a Colorimetric Protein Phosphatase Inhibition Assay and Enzyme Linked Immunosorbent Assay for the Study of Microcystins and Nodularin.* Toxicon, 32: p. 1495-1507.

Bruce, D., P. Westerhoff, and A. Brawley-Chesworth. 2002. *Removal of 2-Methylisoborneol and Geosmin in Surface Water Treatment Plants in Arizona.* AQUA. 51.4: p. 183-197.

Carmichael, W.W. 1992. *Cyanobacteria Secondary Metabolites-Thecyanotoxins.* Journal Applied Bacteriology, 72: p. 6.

Carmichael, W.W. and J. An. 1999. *Using an Enzyme Linked Immunosorbent Assay (ELISA) and Protein Phophatase Inhibition Assay (PP1A) for the Detection of Microcystin and Nodularins.* Natural Toxins, 7: p. 377-385.

Carmichael, W.W. 2002. *Assessment of Blue-Green Algal Toxins in raw and Finished Water,* ed. AWWARF. Denver.

Chorus, I. and J. Bartram. 1999. *Toxic Cyanobacteria in Water: a Guide to Their Public Health Consequences, Monitoring and Management.* UK: E & FN Spon.

De Maagd, P. G., Hendriks, A. J., Seinen, W., and Sijm, D. T. H. M. (1999). "pH-dependent hydrophobicity of the cyanobacteria toxin microcystin-LR." *Water Research*, 33(3), 677-680.

Donati, C., Drikas, M., Hayes, R., and Newcombe, G. (1994). "Microcystin-LR adsorption by powdered activated carbon." *Water Research*, 28, 1735-1742.

Drikas, M., G. Newcombe, and B. Nicholson. 2001. *Water Treatment Options for Cyanobacteria and Their Toxins. in WQTC Proceedings.* AWWA.

EPA. 2002. *Contaminant Candidate List*

Fastner, J., R. Heinze, and I. Chorus. 1995. *Water Science Technology.* 32, (165-170).

Fischer, W. J., Garthwaite, I., Miles, C. O., Ross, K. M., Aggen, J. B., Chamberlin, A. R., Towers, N. R., and Dietrich, D. R. (2001). "Congener-independent immunoassay for microcystins and nodularins." *Environmental Science Technology*, 35(4849-4856).

Hass, C. and Finch, G. 2001. Methodologies for the Determination of Disinfection Effectiveness. In. Foundation, A.R. (ed).

Hayes, K. P., and Burch, M. D. 1989. "Odorous Compounds Associated with Algal Blooms in South Australian Waters." *Wat. Research*, 23, 115-121.

Heresztyn, T. and B.C. Nicholson. 2001. *Determination of Cyanobacterial Hepatotoxins Directly in Water Using a Protein Phosphatase Inhibition Assay.* Water Research. 35: p. 3049-3056.

Honkanen, R. E., Codispoti, B. A., Tse, K., and Boynton, A. L. (1994). "Characterization of natural toxins with inhibitory activity against serine/threonine protein phosphatases." *Toxicon*, 32, 339-350.

Honkanen, R. E., Zwiller, J., Moore, R. E., Daily, S. L., Khatra, B. S., Dukelow, M., and Boynton, A. L. (1990). "Characterization of microcystin-LR, a potent inhibitor of type 1 and type 2A protein phosphatases." *Journal Biol. Chem*, 265, 19401-19404.

Hrudey, S.E., et al. 1993. Liver Toxins and Odour Agents in Cyanobacterial Blooms in Alberta Surface Water Supplies. Disinfection Dilemma: Microbial Control Versus by-Products. In *Fifth National Conference on Drinking Water.* Canada: AWWA.

Hu, Q., M. Sommerfeld, L. Baker, P. Westerhoff. Canal Wall Brushing - a Control Measure for Taste and Odor Problems in Drinking Water Supplies in Arid Environments. *Journal of Water Supply: Research and Technology*. AQUA, 52:8:545-554 (2003)

Izaguirre, G. and W.D. Taylor. 1995. *Geosmin and 2-Methylisoborneol Production in a Major Aqueduct System*. Water Science Technology. 31: p. 41-48.

Izaguirre, G., C.J. Hwang, S.W. Krasner and M.J. McGuire. 1982. Geosmin and 2-methylisoborneol from cyanobacteria in three water supply systems. *Appl. Environ. Microbiol.* 43:708-714.

Jones, G., Burch, M., Falconer, I., and Craig, K. (1993). "Cyanobacterial toxicity. Technical Advisory Group Report, Algal Management Strategy." Murray-Darling Basin Commission, Canberra, Australia.

Khari, D., and al, e. (2000). *Distribution Generated Taste and Odor Phenomena*, AWWARF, Denver, CO.

Lambert, T.W., C.F.B. Holmes, and S.E. Hurley. 1996. *Adsorption of microcystin-LR by activated carbon and removal in full-scale water treatment.* Water Research. 30: p. 1411-1422.

Lloyd, S. W., Lea, J. M., Zimba, P. V., and Grimm, C. C. 1998. Rapid analysis of geosmin and 2-methylisoborneol in water using solid phase micro extraction procedures. *Water Research*, 32(7), 2140-2146.

Maier, R.M., I.L. Pepper, and C.P. Gerba. 2000. *Environmental Microbiology*. Academic Press.

Means, E.G. III, and M.J. McGuire. 1986. An early warning system for taste and odor control. *Journal AWWA*, March, 77-83.

Meng, A.K., L. Brenner, and I.H. Suffet. 1992. Correlation of chemical and sensory data by principal component factor-analysis. *Water Science and Technology* 25 (2): 49-56.

Miller, M. J., Critchely, M. M., Hutson, J., and Fallowfield, J. J. (2001). "The adsorption of cyanobacterial hepatotxins from water onto soil during batch experiments." *Water Research*, 35, 1735-1742.

Mohammed, Z. A., Carmichael, W. W., An, J., and El-Sharouny, H. M. (1999). *Activated carbon removal efficiency of microcystins in an aqueous cell extract of Microsystis aeruginosa and Oscillatorai tenuis strains isolated from Egyptian freshwaters*, John Wiley and Sons.

Newcombe, G., Dook, D., Brooke, S., and Ho, L. "Mirocystin-LR and LA similarities and differences in response to water treatment processes." *WQTC Proceedings*.

Nguyen, M., L.A. Baker, and P. Westerhoff. 2002. *DOC and DBP Precursors in Western US Watersheds and Reservoirs*. JAWWA. 94: p. 98-112.

Nicholson, B.C., J. Rositano, and M.D. Burch. 1994. *Destruction of Cyanobacterial Peptide Hepatotxins by Chlorine and Chloramines.* Water Research, 28: p. 1297-1303.

Nicholson, B. and G. Shaw. 2001. *Congener-Independent Immunoassay for Microcystins and Nodularins.* Environmental Science Technology. 35: p. 4849-4856.

Nicholson, B. and G. Shaw. 2001. Instrumental Methods for the Determination of Cyanobacterial Toxins. In *WQTC Proceedings*. AWWA.

Pouria, S., de Andarade, A., Barbosa, J., Cavalcanti, R. L., Garetto, V. T. S., Ward, C. J., Preiser, W., Poon, G. K., Neild, G. H., and Codd, G. A. (1998). "Fatal microcystin intoxication in haemodialysis unit in Caruaru, Brazil." *Lancet*, 352(21-26).

Rapala, J., Sivonen, K., Luukkainen, R., and Niemela, S. I. (1993). "Anatoxin-a concentration in *Anabaena* and *Aphanizomenon* under different environmental conditions and comparisons of growth by toxic and non-toxic *Anabaena* strains-a laboratory study." *Journal Applied Phycology*, 5(6), 581-591.

Rivasseau, C., Racaud, P., Deguin, A., and Hennion, M. C. (1999). "Evaluation of an ELISA kit for the monitoring of microcystins (cyanobacterial toxins) in water and algae environmental samples." *Environmental Science Technology*, 33, 1520-1527.

Skulberg, O. M., Carmichael, W. W., Codd, G. A., and Skulberg, R. (1993). *Taxonomy of toxic cyanophyceae (cyanobacteria)*, London.

Sommerfeld, M., P. Westerhoff, and L. Baker. 2002. *Reducing Taste and Odor and Other Algae-Related Problems From Surface Water Supplies in Arid Environment.* Arizona State University, City of Phoenix Final Report.

Stal, L.R. 1995. *Physiological Ecology of Cyanobacteria in Microbial Mats and Other Communities.* New Phytol.

Suffet, I.H., C. Anselme, and J. Mallevialle. 1993. AWWA Taste and Odor Survey. In *AWWA Water Quality and Technology Conference.* Miami, Florida.

Suffet, I.H., J. Mallevialle, and E. Kawcynski. 1995. Advances in taste and odor control. In *AWWARF Final Report.* Denver, Colorado.

Ueno, Y., Nagata, S., Tsutsumi, T., Hasegawa, A., Watanabe, M. F., Park, H. D., Chen, G. C., Chen, G., and Yu, S. Z. (1996). "Detection of microcystins, a blue-green algal hepatotoxin, in drinking water sampled in Haimen and Fusui, endemic areas of primary liver cancer in China, by highly sensitive immunoassay." *Carcinogenesis*, 17(6), 1317-1321.

Watson, S. B., Brownlee, B., Satchwill, T., and Hargesheimer, E. E. 2000. Quantitative Analysis of Trace Levels of Geosmin and MIB in Source and Drinking Water Using Headspace SPME. *Water Research*, 34(10), 2818-2828.

WHO. 1993. *Guidelines for Drinking-Water Quality.* World Health Organization 2nd Edition. Vol. 1. Geneva.

Wnorowski, A.U. 1992. *Tastes and Odors in the Aquatic Environment: a Review.* Water SA. 18: p. 203-214.

Wolfram, S. 2002. Mathematica 4 for Students. In. Champaign, IL: Wolfram Research.

Yoo, R. S., Carmichael, W. W., Hoehn, R. C., and Hrudey, S. E. (1995). *Cyanobacterial (Blue-green algal) toxins: A Resource Guide*, AWWARF, Denver, CO.

Young, C. C., Suffet, I. H., Crozes, G., and Bruchet, A. (1999). "Identification of a woody-hay odor-causing compound in a drinking water supply." *Water Science and Technology*, 40(6), 273-278.

ABBREVIATIONS

AWWA	American Water Works Association
AwwaRF	Awwa Research Foundation
° C	degrees Celsius
CAP	Central Arizona Project
CFU	colony forming units
COP	City of Phoenix
CT	contact time
DOM	dissolved organic matter
ELISA	Enzyme Linked Immunorsorbent Assay
FPA	Flavor Profile Analysis
GC/MS	Gas Chromatograph/Mass Spectrometry
HPLC	High Performance Legend Chromatography
MC	Microcystin
MDL	method detection limit
MeOH	methanol
MIB	2-Methylisobrneol
PAC	powder activated carbon
PBS	phosphate buffer solution
POC	particulate organic carbon
PP2A	Protein Phosphates - 2A Assay
QC	quality control

SCADA	Supervisory Control and Data Acquisition
SPME	Solid Phase Micro Extraction
SRP	Salt River Project
T&O	taste and odor
TDS	total dissolved solids
WHO	World Health Organization
WTP	water treatment plant
YMEA	yeast-malt-extract agar

IWA Publishing's authorised EU representative for General Product Safety Regulations is Diane D'Arras, 15 rue Duret, 75116 Paris, France, e-mail: safety@iwap.co.uk.

Printed and bound by CPI Group (UK) Ltd, Croydon, CR0 4YY
17/09/2026
02218808-0017